Advance praise for
Eating Fossil Fuels

People eat — and this book explains in the most lucid way *what* they eat: namely the product of energy-intensive agriculture. It finds that the Green Revolution, hailed as a breakthrough by which to feed an exploding population, actually degrades the ecosystem, making us ever more dependent on energy inputs from oil and gas. But oil and gas are set to deplete to near exhaustion this century. The challenges are great, but there are solutions. This book is key reading for those wanting to be counted amongst the survivors.

— C. J. Campbell, Chairman, The Association for the Study of Peak Oil (ASPO)

In retrospect, the industrialization of agriculture was one of the greatest blunders in the history of our species; as Dale Allen Pfeiffer shows, it is a mistake that can be undone — and, given the imminent peak in global oil production, *must be* undone as soon as possible. The world's addiction to oil is as personal and prosaic as what's on your dinner plate and how it got there. This is a book of enormous importance.

. — Richard Heinberg, author of *The Party's Over, Powerdown,* and *The Oil Depletion Protocol*

Eating Fossil Fuels is a wake-up call for humanity. It traces how, with industrialization and globalization, we have stopped eating real food and have started to eat oil, increasing the fossil fuel content of the food chain and threatening the environment, our health and our future. Pfeiffer shows how creating fossil fuel-free ecological and localized food systems has become a central challenge for sustainability, and how you can help make this shift.

— Dr. Vandana Shiva, author of *Earth Democracy: Justice, Sustainability,* and *Peace and Stolen Harvest: The Hijacking of the Global Food Supply*

D1052491

As civilization stumbles blindly into the Post-Carbon Age, Dale Allen Pfeiffer has astutely identified our most critical vulnerability: the intimate dependency of our food system on finite fossil fuels. His book also includes a plethora of solutions and resources to help guide us to a sustainable, low-energy future.

— John G. Howe, author of *The End of Fossil Energy and The Last Chance for Survival*

Eating Fossil Fuels is vital for those concerned about peak oil and climate change who are looking both for an understanding of fossil fuels in food production and an action plan.

— E. R. "Pat" Murphy, Executive Director, The Community Solution and Producer of the Film "The Power of Community: How Cuba Survived Peak Oil"

EATING FOSSIL FUELS

EATING FOSSIL FUELS

Oil, Food and the Coming Crisis in Agriculture

Dale Allen Pfeiffer

NEW SOCIETY PUBLISHERS

CATALOGING IN PUBLICATION DATA:
A catalog record for this publication is available
from the National Library of Canada.

Copyright © 2006 by Dale Allen Pfeiffer.
All rights reserved.

Cover design by Diane McIntosh.
Photos: iStock Photo and Alamy Images.

Printed in Canada.
Second printing August 2008.

Paperback ISBN 13: 978-0-86571-565-3
Paperback ISBN 10: 0-86571-565-3

Inquiries regarding requests to reprint all or part of *Eating Fossil Fuels*
should be addressed to New Society Publishers at the address below.

To order directly from the publishers, please call toll-free
(North America) 1-800-567-6772, or order online at
www.newsociety.com

Any other inquiries can be directed by mail to:
New Society Publishers
P.O. Box 189, Gabriola Island, BC V0R 1X0, Canada
1-800-567-6772

New Society Publishers' mission is to publish books that contribute in
fundamental ways to building an ecologically sustainable and just society, and to do so with the least possible impact on the environment, in a
manner that models this vision. We are committed to doing this not just
through education, but through action. We are acting on our commit-
ment to the world's remaining ancient forests by phasing out our paper
supply from ancient forests worldwide. This book is one step toward
ending global deforestation and climate change. It is printed on acid-free
paper that is 100% old growth forest-free (100% post-consumer recycled), processed chlorine free, and printed with vegetable-based, low-
VOC inks. For further information, or to browse our full list of books
and purchase securely, visit our website at: www.newsociety.com

NEW SOCIETY PUBLISHERS   www.newsociety.com

To my loving wife Elizabeth Anne Pfeiffer.

Without your support
I would have given up on this long ago.
And without your positive input,
I would have drowned in pessimism.

Contents

Acknowledgments

I wish to acknowledge all of the scientists, engineers and researchers whose work I draw upon in this book, specifically David Pimentel, Colin J. Campbell, Jean Laherrère, Richard Duncan, Walter Youngquist, Mario Giampietro, Sandra L. Postal, Richard Pirog, Tony Boys, David Von Hippel, Peter Hayes, M. Sinclair and M. Thompson. I would also like to thank John T. Heinen and Beth Baker for all of their help, including acting as the agent for this book. Without your work, this book would not have been possible.

Foreword

Π WAS FORTUNATE in meeting Dale Allen Pfeifer this spring at
a Peak Oil Conference in New York: Local Energy Solutions. I re-
call being struck by his factual, measured presentation and his
analysis of key responses required in order to meet the challenges
we face as a global community. A key theme was local responses
and Dale presented a range of appropriate local responses perti-
nent to the general themes of food, energy, shelter, water and
economy: themes that are the core focus of this book.

Subsequently, I learned of Dale's long familiarity with the subject of Peak Oil and his work to alert the general public to its far-
reaching implications. It is heartening to me that we find much
common ground in our assessment of needed response. He has
taught me much about the realities of geology and exploration,
and where this places us in terms of our fossil energy future.
Having seen U.S. military hardware protecting the Oil Ministry
building in Baghdad in 2003 while other ministries were bull-
dozing the remains of burned and looted files from their parking
lots, I have some appreciation for the petrochemical influence on
foreign policy processes and the human implications of resultant
actions.

I approach the current situation from a background in ecology
and over a decade of work in the broad field of international disaster response and 'development.' My first international aid-related

work was in northern Iraq between 1992 and 1995. My observations of the patterns of aid response from the ground through the 1990s led me, in 1999, to seek out a system that would integrate people, environment and design in a holistic manner. I was concerned that much of what we did mirrored unsustainable systems derived from the cultures of countries that provided the 'solutions,' or was simply lacking in an integrated design perspective. It was in this way that I came to do my first permaculture design course: the 72-hour intensive course based upon a curriculum developed by permaculture founder, Bill Mollison, first taught in the early 1980s and subsequently enriched by the work of many.

Permaculture is a contraction of two words: **perma**nence and agri**culture**, or culture as there is no true culture without permanent agriculture. It is a design science that uses natural systems as the model for creating productive systems with the resilience, diversity and stability of natural ecosystems. Based on the foundational ethics of earth care, people care and return of surplus (fair share), permaculture works as a linking science with a set of core principles derived from nature. The outcome of good design is to minimize our footprint through efficient and harmonious use of resources in the creation of systems that are mutually supportive of key functions or needs. It rests firmly in an understanding of the energy equations highlighted within this book as a basis for design.

A sustainable system is one which, over its lifetime, produces the energy required to develop and maintain itself. In an agricultural context, this means that the system must be actively building soil and 'ecosystemic' (supporting functional ecosystem services). Broad-scale industrial agriculture as popularly practiced today has moved out of balance in terms of an energy audit, environmental impacts and the quality of the products as shown in chapters 1–4 of this book. Whenever we work outside of a natural scale, our systems become energetically chaotic and damaging. Likewise, as we harmonize with natural patterns, we are able to return to abundance.

Polycultural (mixed) production systems can easily be designed to produce 3–10 lb of food per square foot: many times that gained by current industrial agriculture, but with increased need for human involvement as designers and workers and partnering with biological systems. This can be achieved by shaping land to maximize natural water function and storage in the soil, using legume (nitrogen fixing) plants and trees within the system, and majoring in trees and perennial plants as the main productive crops rather than annual crops. The 'secret' to doing this is not in any one thing but in a beneficial linking of functional elements within an overall design: stacking plant functions in time and space, using combinations of plants that are mutually beneficial. It involves actively promoting healthy, balanced life within the soil by cover cropping, using mulch and generally exposing soil as little as possible whilst allowing on-site organic waste to return to the soil cycle.

As our global population reaches 50 percent urban, there is an urgent need to return food, fiber and materials production to the cities. As Dale has clearly pointed out, the U.S. urban population (and many in more rural settings) is currently highly food insecure due to a lack of local food production and networks. Solutions for cities include the return of organics and humanure to bioregional growing systems, increasing tree and perennial crops within open and green space, using vertical space on building structures for growing systems and climbing plants, and understanding how to work with microclimates that can be created in the unique urban environment. Water can be directed to ponds and growing systems rather than the traditional manner of running it out of the system using the shortest path.

Permaculture teaches us that our first task is to take responsibility for ourselves and those we love, and the way in which we provide for our needs. Becoming actively and positively engaged in meeting our needs and linking with or developing a broader community is a natural outcome of such awareness. The tools provided in the last chapter of this book provide an appropriate place to begin. Just as the cells in our bodies have banded together to get smarter by sharing awareness of environmental and other factors and cooperating in response (Bruce Lipton, 2005, *The Biology of Belief*), so our challenge on the macro level is to do the same.

It matters not what we call what we do so long as it conforms to ecological patterns, works with human communities and is applied, shared and replicated where relevant. That is our urgent task and one which we may relish as we share and apply proven solutions that work. Such sharing may occur using bioregional models for organizing resources in sensible patterns in relation to place and those who live there — and applying broader climatic design models across areas of similar climate. In this manner, we are encouraged to work locally and share our models globally in areas where they are relevant.

Andrew Jones
Baja California Sur
July 2006

Andrew Jones works as a sustainable designer, project implementer and teacher using a permaculture framework. Holding a degree in Ecology and postgraduate studies in Environmental Management and Development, he has worked in an international aid and development context since 1992, including sustainable business initiatives in the US since 2001. Currently based in Brooklyn, NY, he is a board member of the Permaculture Research Institute and involved in a range of international permaculture initiatives.

Introduction

THE GREEN REVOLUTION, which began in the 1960s, did not cure world hunger. But it did transform food production into an industry, and allow for the consolidation of small farms into what have become the large agribiz corporations. While the abundance of cheap food that resulted did nothing to alleviate world hunger, it did allow the human population to grow far in excess of the planet's carrying capacity.

The Green Revolution achieved all of this by making food production extremely dependent on fossil fuels. The globalization of food production during the 1980s, 1990s, and on into the new millennium has finished the job of demolishing localized agriculture. Globalization has given us access to exotic foods and crops that may be out of season in our locale, but it has done so by increasing the vulnerability of food security. If global food shipments were to stop tomorrow, we would no longer be able to feed ourselves.

The Green Revolution and the globalization of food production were fostered by the availability of cheap, abundant hydrocarbon energy in the form of oil and natural gas. The fertilizers we feed our crops are generated from natural gas, and the pesticides that protect our monoculture crops from devastating infestations are derived from oil. We are dependent on the energy of oil and natural gas to seed our crops, maintain them, harvest them, process them, and transport them to market.

The intensive practices of industrialized agriculture quickly strip the soil of nutrients and deplete easily accessible water supplies. As a result, the need for hydrocarbon-based inputs must increase, along with increasingly energy-intensive irrigation. Without hydrocarbons, much of the world's farmlands would quickly become unproductive.

Yet hydrocarbons are a nonrenewable resource, and growing evidence indicates that world hydrocarbon production will peak around 2010, followed by an irreversible decline. The impact on our agricultural system could be catastrophic. As the cost of hydrocarbon production increases, food could be priced out of the reach of the majority of our population. Hunger could become commonplace in every corner of the world, including your own neighborhood.

The solution is to relocalize agriculture. We need to rebuild our local food production infrastructure. Agribiz corporations are not going to do this, and their client governments refuse to recognize the problem. It is up to us to resuscitate localized agriculture through the development of a grassroots movement. This book will give you a glimpse of the efforts needed to relocalize food production, and will provide you with contacts for cooperatives and organizations in your area that are already working towards this goal.

If we can build a grassroots relocalized agriculture movement, then we may be able to cushion ourselves against the coming decline of hydrocarbon production. Given a sustainable agriculture, our population would be able to decline with a minimum of hardship until our numbers are below the carrying capacity of the planet.

In that regard, we are talking about the ultimate diet plan, and this is the ultimate diet book.

Dale Allen Pfeiffer
Irvine, KY Appalachia
March 2006

1

Food = Energy + Nutrients

EVERYTHING LIVING DEPENDS on energy, which is replenished through some food source. Without energy, we would literally run down until our bodies failed. Plants take their energy directly from the sun, via the miracle of photosynthesis. Plant cells use chlorophyll to turn the sun's energy into carbohydrates, which store that energy in molecular bonds that can then be used to do biological work. Animals, in turn, must feed on plants or on other animals that have, themselves, fed on plants. And decomposers like bacteria and fungi feed on the detritus and dead bodies of animals and plants. In a way, you could say that this whole wonderfully complex biosphere with its intricate web of relationships is simply a circulatory system for solar energy.

Millions of years ago, around the time of the dinosaurs, during a period of global warming, the sun's energy gave rise to great algal mats in the oceans. These were huge colonies of single-celled plant life dedicated to self-replication and the conversion of solar energy to carbohydrates. Over the centuries, dead algae rained down on the ocean floor, where they built up a layer of organic detritus. Eventually this organic ooze was buried under inorganic sediment, and there it was compacted and heated over long periods of time, until the carbohydrates were eventually transformed into what we now call fossil fuels.

3

Where the detritus was subjected to temperatures above a certain threshold, or where it was mixed with the remains of land vegetation washed in from coastal areas, it became natural gas. And where the detritus was pure and remained below the critical temperature, it formed complex hydrocarbon molecules of concentrated energy content, which we now call crude oil. The hydrocarbon resources of this planet are actually a deposit of solar energy made millions of years ago and processed over the passage of time.

Over time, the climate altered and the algal blooms receded. The age of the dinosaurs ended, to be replaced by the age of the mammals. And eventually one branch of mammals evolved a two-legged creature capable of abstract thought and the conscious manipulation of its own environment. Human beings developed not as the ultimate result of life on this planet, but as just another permutation of a biosphere given to creating ever-more complex pathways for drawing out the process of entropy.

ENTROPY, LIFE AND FOSSIL FUELS

Entropy is a measure of the dispersal of energy. It measures how much energy is spread out in a particular process, or how widely spread out it becomes. When energy is diffused, it is unavailable to do useful work. So, the higher the entropy, the less energy is available to do useful work.

All physical systems move from a state of low entropy to a state of high entropy. The amount of energy *available* in a system is always less than the total energy of the system. Whenever energy changes forms, or is used, a portion of it is lost to entropy.

It is important to understand that the total amount of entropy in the universe is always growing, and can never be diminished. We can maintain the appearance of reducing entropy in a subsystem only by bringing in energy from outside of that subsystem and exporting entropy. But the total entropy of the universe will only increase.

Living things engage in a sort of shell game with regard to entropy, by hiding their entropy production outside of their subsystem. But in the end, they are really performing a balancing act because they have not reduced entropy, only shifted it elsewhere. Life requires low entropy and cannot exist in a high entropy environment. Let us look at the brewing process to illustrate the relationship between life and entropy.

A brewer's vat full of mash is a low entropy environment rich in carbohydrates and sugars. When we introduce yeast to this vat, they will begin eating and multiplying. The

In the beginning, these human beings were hunters and gatherers. They existed, as did all other animal life on this planet, by foraging plant food or by hunting other animals. Then, about 10,000 years ago, human beings made what is considered to be the most significant change in their history, the invention of agriculture. By planting, tending, and harvesting crops on one piece of land, human beings were able to produce more food than by simply foraging through their territory. Farming gave them enough food to ensure that they would make it through the year without starving. It gave them food security. The advent of agriculture prepared the way for civilization as we know it.

Through agriculture, human beings were able to process and harvest solar energy. The sudden abundance of agricultural produce led inevitably to population growth. The growing population, in turn, required an expansion of agriculture. And so humanity began its first population explosion. This growth came at the expense of other life forms, which were pushed out of their

growing population of yeast produces high entropy in the form of carbon dioxide and ethanol. When the vat exceeds some critical level of entropy, the yeast will die off. Some yeast will remain to feed on the little remaining low entropy, but the vat will never return to its low entropy state without being emptied and refilled.

Human beings have taken the creation of entropy to new levels. It would seem that the one thing our socio-economic system does exceedingly well is to produce entropy. The human brewing vat (and our civilization) is currently subsidized by abundant, cheap fossil fuels. Human beings grow on this mash just as did the yeast in our example, multiplying our numbers and producing an abundance of consumer goods. In addition, we have produced high entropy in the form of environmental degradation, garbage, pollution and global warming. Now we are approaching the critical level of entropy that will result in a die-off.

To avoid reaching this critical level of entropy, we need to slow the production of entropy below the level of incipient solar energy. To do this, we must abandon the dominant socio-economic system based on constant growth and consumption. If we exceed the critical level of entropy, we will experience a die-off just as the yeast did. Furthermore, as there is no way to reduce entropy, we will never recover from this die-off.

habitat by farming and civilization. Photosynthesis sets a limit on the amount of food that can be generated at any one time, and therefore places a limit on population growth. Solar energy has a limited rate of flow into this planet. For our purposes, the amount of incoming solar energy is a constant. And only a portion of this solar energy is captured through the process of photosynthesis.

To increase food production, these early farmers had to increase the acreage under cultivation and displace their competitors. There was no other way to increase the amount of energy available for food production. Human population grew by displacing everything else and appropriating more and more of the available solar energy.

The need to expand agricultural production has been one of the root causes behind most of the wars in recorded history. When Europeans could no longer expand cultivation, they began the task of conquering the world. Explorers were followed by conquistadors and traders and settlers. The declared reasons for this expansion may have been trade, avarice, empire, or simply curiosity, but at the base it was all about the expansion of agricultural productivity. Wherever explorers and conquistadors traveled, they may have carried off loot, but they left plantations. Settlers toiled to clear land and establish their own homesteads. This conquest and expansion went on until there was no place left for further expansion. Certainly, to this day landowners and farmers fight to claim still more land for agricultural productivity, but they are fighting over crumbs. Today, virtually all the productive land on this planet is being exploited by agriculture. What remains is either too steep, too wet, too dry, or lacking in soil nutrients.[1]

Just when output could expand no more by increasing acreage, agricultural innovations made possible a more thorough exploitation of the acreage already available. The process of "pest" displacement and land appropriation for agriculture accelerated with the industrial revolution as the mechanization of agriculture sped up the clearing and tilling of land and increased the amount of farmland that could be tended by one person. And with every increase in food production, the human population grew apace.

The amount of solar energy that can actually be harnessed through photosynthesis is called the photosynthetic capability. This is further limited by geography, climate and soil type. The photosynthetic capability of a field with rich soil, for example, is high compared to the photosynthetic capability of a rocky mountain slope. At present, nearly 40 percent of all land-based photosynthetic capability has been appropriated by human beings.[2] In the United States, agriculture diverts more than half of the energy captured by photosynthesis.[3] We have taken over all the prime real estate on this planet. The rest of the biota is forced to make due with what is left. Plainly, this is one of the major factors in species extinctions and ecosystem stress.

The Green Revolution

At just the point when agriculture was running out of unexploited tillable lands, technological breakthroughs in the 1950s and 1960s allowed it to continue increasing production through the use of marginal and depleted lands. This transformation is known as the Green Revolution. The Green Revolution resulted in the industrialization of agriculture. Part of the advance resulted from new hybrid food plants, leading to more productive food crops. Between 1950 and 1984, as the Green Revolution transformed agriculture around the globe, world grain production increased by 250 percent.[4] That is a tremendous increase in the amount of food energy available for human consumption. This additional energy did not come from an increase in sunlight, nor did it result from introducing agriculture to new vistas of land. The energy for the Green Revolution was provided by fossil fuels. The Green Revolution was made possible by fossil fuel-based fertilizers and pesticides, and hydrocarbon-fueled irrigation.

The Green Revolution increased the energy flow to agriculture by an average of 50 times its traditional energy input.[5] In the most extreme cases, energy consumption by agriculture has increased a hundredfold or more.[6] In a very real sense, we are eating fossil fuels.

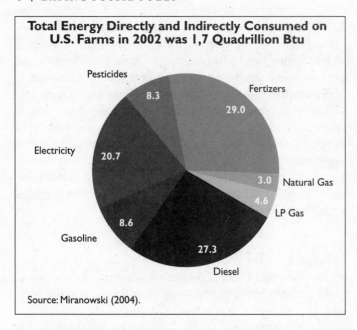

Total Energy Directly and Indirectly Consumed on U.S. Farms in 2002 was 1,7 Quadrillion Btu

Pesticides 8.3
Fertizers 29.0
Electricity 20.7
Natural Gas 3.0
LP Gas 4.6
Gasoline 8.6
Diesel 27.3

Source: Miranowski (2004).

In the United States, the equivalent of 400 gallons of oil is expended annually to feed each US citizen (as of 1994).[7] US agricultural consumption of energy by source is broken down in this pie chart.[8]

A report of energy usage in Canadian farms came up with the following break down:

- 31 percent for the manufacture of inorganic fertilizer
- 19 percent for the operation of field machinery
- 16 percent for transportation
- 13 percent for irrigation
- 8 percent for raising livestock (not including livestock feed)
- 5 percent for crop drying
- 5 percent for pesticide production
- 3 percent miscellaneous[9]

In addition, the energy costs of processing, distributing, and preparing food in the home far exceed the energy used on the farm to produce it.[10]

Failure of the Green Revolution

Due to the laws of thermodynamics, there is not a direct correspondence between energy inflow and outflow in agriculture. Along the way, there is a marked energy loss. Between 1945 and 1994 energy input to agriculture increased fourfold while crop yields only increased threefold.[11] Since then, energy input has continued to increase without a corresponding increase in crop yield. We have reached the point of marginal returns. Yet, due to soil degradation, the increased demands of pest management, and increasing energy costs for irrigation (all of which is examined below), modern agriculture must continue increasing its energy expenditures simply to maintain current crop yields. The Green Revolution is becoming bankrupt.

In the 1950s and 1960s, the Green Revolution was promoted as the solution to world hunger. The rise in agricultural production was supposed to ensure that there was enough food for everyone and that no one on the planet would go hungry. Unfortunately, this has not been the case.

FIRST LAW OF THERMODYNAMICS or THE ENERGY CONSERVATION LAW
Energy can neither be created or destroyed;
or
The energy of the universe remains constant;
or
You can't win.

SECOND LAW OF THERMODYNAMICS or THE ENTROPY LAW
Without compensating changes elsewhere, heat can flow only from a hotter to a colder body;
or
With passing chronological time, the entropy of the universe tends towards a maximum;
or
You can't break even.

THIRD LAW OF THERMODYNAMICS
The entropy of any condensed substance, i.e., liquid or solid, has at zero absolute temperature the value zero;
or
Zero absolute temperature cannot be reached;
or
You have to stay in the game.

In spite of a 70 percent population increase, the Green Revolution has led to a 17 percent increase in calories available per person. Everyone in the world could have a daily intake of at least 2,720 kilocalories (1 kilocalorie = 1,000 calories) if food were distributed more equitably.[12] Yet, there were still an estimated 798 million undernourished people in developing countries as of 1999–2000. This is a decrease of only 19 million from the 1990–1992 estimate.[13]

In the United States, 11.9 percent of all households are considered to be "food insecure" — they are not sure if they will have a meal from one day to the next. For single-mother families, this figure soars to 31.7 percent. For the working poor it is 35 percent. Among American adults, the rate is 10.8 percent, while for children it is nearly double, at 18.2 percent. These percentages translate into nearly 35 million people, 13 million of them children, living hungry or on the edge of hunger. From 1999 to 2002, the number of hungry people in the US increased by 3.9 million, more than 1 million of them children.[14]

The failure of the Green Revolution results from a misunderstanding of the causes of starvation in the world today. Hunger is not caused from lack of food, but from a lack of access to food. It is the failure of the profit-based market to distribute food equitably. Those who can afford food have a diet far in excess of their needs, while those who cannot afford food go hungry. The key to ending world hunger lies in social reforms, agrarian reforms, and true democratic reforms, along with the recognition that everyone has a right to a subsistence diet.[15]

The Green Revolution has, in fact, contributed to the inequity of our agricultural system by making it more difficult for farmers to compete with agricultural corporations. Not only does the Green Revolution fail in its stated mission, its industrial practices lead to degradation of the land and water supply. In the end, Green Revolution-type agriculture is unsustainable.

2

Land Degradation

LAND DEGRADATION — mostly due to soil erosion, mineral depletion, and urbanization — is becoming a major problem worldwide. Currently, the full scope of the problem is hidden because lost vitality is compensated for by intensified use and increased application of artificial fertilizers. In the end, these strategies only exacerbate the problem, leading to a total collapse of soil viability.

Since 1945, the total land degraded by soil depletion, desertification, and the destruction of tropical rainforests comes to more than 5 billion hectares, or greater than 43 percent of the Earth's vegetated surface.[1]

Each year, 10 million hectares of productive, arable land are abandoned due to severe degradation.[2] At the same time, 5 million hectares must be added to production to feed the extra 84 million humans born each year. In all, 15 million hectares are needed yearly to make up for losses and increased population. Most of this land is coming from the forests,[3] accounting for 60 percent of world deforestation.[4]

It takes 500 years for nature to replace 1 inch of topsoil.[5] Approximately 3,000 years are needed for natural reformation of topsoil to the depth needed for satisfactory crop production.[6] In a natural environment, topsoil is built up by decaying plant matter and weathering rock, and it is protected from erosion by growing plants. In soil made vulnerable by agriculture, erosion is reducing productivity up to 65 percent each year.[7] Former prairie lands, which constitute the breadbasket of the United States, have lost one half of their topsoil after being farmed for about 100 years. This soil is eroding 30 times faster than the natural formation rate.[8] Food crops are much hungrier than the natural grasses which once covered the Great Plains. As a result, the remaining topsoil is increasingly depleted of nutrients. Soil erosion and mineral depletion remove about $20 billion worth of plant nutrients from US agricultural soils every year.[9] Much of the soil in the Great Plains is now little more than a sponge into which we must pour hydrocarbon-based fertilizers in order to produce crops.

Erosion rates are increasing throughout the world. China is losing topsoil at the rate of 40 tons per hectare per year (t/ha/yr).[10] During the spring planting season, Chinese soil can be detected in the atmosphere as far away as Hawaii.[11] China's need for food exceeded the capacity of its agricultural system decades ago. The Chinese have pushed their agricultural lands to the very limit, using artificial fertilizers and pesticides to force every bit of production from the land. And still they must import prodigious quantities of grain from the US and other countries in order to feed their population.[12]

Agricultural production in some parts of Africa has declined by 50 percent due to soil erosion and desertification.[13] As with China, soil eroded from Africa can be detected in Florida and Brazil.[14] Serious production losses (in the 20 percent range) have also been reported for India, Pakistan, Nepal, Iran, Jordan, Lebanon, and Israel.[15] Globally, the loss of 75 billion tons of soil per year costs the world about $400 billion annually, or about $70 per person per year.[16]

In the US, assuming erosion by wind and water at a rate of 17 t/ha/yr, the onsite costs of irrigation to replace lost soil moisture and fertilizers to replace lost nutrients translates into approximately $28 billion per year, in 1992 dollars.[17] In addition to this, there are the offsite costs of soil erosion: roadway, sewer, and basement siltation; drainage disruption; foundation and pavement undermining; gullying of roads; earth dam failures; eutrophication of waterways; siltation of harbors and channels; loss of reservoir storage; loss of wildlife habitat and disruption of stream ecology; flooding; damage to public health; and increased water treatment costs. Offsite costs are estimated at $17 billion per year, also in 1992 dollars.[18] The combined cost is $45 billion per year, or about $100/hectare of pasture and farmland, which increases the production costs of US agriculture by 25 percent.[19]

Approximately three-quarters of the land in the United States is devoted to agriculture and commercial forestry.[20] Of this, every year more than two million acres of cropland are lost to erosion, salinization, and water logging. On top of this, farmland loses another million acres to urbanization, road building, and industry annually.[21] Between 1945 and 1978, an area equivalent to the states of Ohio and Pennsylvania was blacktopped. Of the land taken for housing and highways, almost half was among the most agriculturally productive land in the country.[22] Incidentally, only a small portion of US land area remains available for the solar energy technologies necessary to support a solar energy-based economy. The land area for harvesting biomass is likewise limited. For this reason, the development of solar energy or biomass must be at the expense of agriculture.

This is the cost of poor farm management. Soil degradation is caused primarily by loss of vegetation cover. It can be combated by a number of techniques, including no-till agriculture, contouring, cover cropping, crop rotation, contour strip cropping, contour buffer planting, terracing, grassed waterways, farm ponds, check dams, and reforestation. Instead, we presently make up for the loss in soil fertility with artificial fertilizers derived from natural gas. It is to be wondered why we bother planting crops in the ground at all — as opposed to growing them with hydroponics — except that the ground provides a cheap and abundant medium.

3

Water Degradation

OVER 70 PERCENT of the Earth's surface is covered by water. However, 97.5 percent of that is salt water, leaving only 2.5 percent as fresh water. Of that fresh water, over 70 percent is frozen in the continental glaciers of Antarctica and Greenland. Most of the remainder is soil moisture, or deeply buried in inaccessible aquifers. Only 0.77 percent of all fresh water — or less than 0.007 percent of all the water on the Earth — is available for human use.[1] This is the water found in rivers and lakes, and groundwater shallow enough to be accessed economically.

Modern agriculture places a severe strain on our water resources. In the US, for example, it consumes fully 85 percent of all freshwater resources.[2]

In the last century, irrigated land area increased nearly five-fold. Although only 17 percent of the world's cropland is irrigated, that 17 percent produces 40 percent of the global harvest.[3] This disproportionate share is largely due to the capability of irrigated lands to produce two and sometimes three crops in a year. However, environmental damage to irrigated cropland from salinization and waterlogging, in combination with rising irrigation costs and other factors, has slowed the increase in irrigated land in the last few decades so that it now no longer keeps up with population growth. Per capita irrigated area peaked in 1978 at 0.48 hectares per person.[4] Since then, it has fallen more than seven percent.[5]

Poor irrigation practices have led to the steady accumulation of salt in soil, damaging 60 million hectares, or about one quarter of the world's irrigated land.[6]

Overdraft is occurring in many surface water resources, especially in the west and south. The typical example is the Colorado River, which is reduced to a trickle by the time it reaches the Pacific.[7] The lower reaches of China's Yellow River have gone dry for an average of 70 days a year in each of the last 10 years; in 1995 the river was dry for 122 days.[8] Similar stories can be told about the Ganges, the Nile, and many other rivers around the world.[9] During the dry season, essentially no water is released to the sea from much of Asia. In the former Soviet Union region of Central Asia, the Aral Sea has lost half of its area and two-thirds of its volume, due to river diversions for cotton production. Tripled salinity levels in the Aral Sea have wiped out all of the native fish and devastated the local fishing industry.[10]

Surface water supplies only 60 percent of the water used in irrigation. The other 40 percent, and in some places the majority of irrigation, comes from underground aquifers that are pumped so far in excess of their recharge rates as to be, for all intents and purposes, nonrenewable resources. Groundwater is recharged slowly by the percolation of rainwater through the Earth's crust. Less than 0.1 percent of the stored ground water mined annually is replaced by rainfall.[11]

Water tables are dropping a meter or more each year beneath a large area of irrigated farmland in north China; they are falling 20 centimeters a year across two-thirds of India's Punjab, that nation's breadbasket.[12] One-fifth of irrigated land in the US is watered by pumping in excess of the recharge rate. The Southwest receives only 6 percent of the country's available water as rainfall, but its large irrigated farms and growing urban areas account for 36 percent of the nation's water use. California also consumes more water annually than the state receives in rainfall.[13] In Texas, where groundwater depletion is particularly severe, the amount of irrigated land has fallen by more than 30 percent from its peak in 1974.[14]

The Ogallala Aquifer that supplies agriculture, industry and home use in much of the southern and central plains states has an annual overdraft 130 to 160 percent in excess of replacement. This vitally important aquifer will become unproductive in another thirty years or so.[15] The Ogallala Aquifer is the irrigation source for much of the American breadbasket; when it becomes unproductive, the US heartland will go dry. There is talk of diverting water from the Mississippi or the Great Lakes, but these projects pose considerable engineering problems. They would also damage other industries, and are opposed by the states bordering both waterways (and Canada in the case of the Great Lakes). Moreover, such diversions would be extremely energy-dependent.

We can illustrate the demand that modern agriculture places on water resources by looking at a farmland producing corn. A corn crop that produces 118 bushels/acre/year requires more than 500,000 gallons/acre of water during the growing season. The production of 1 pound of maize requires 1,400 pounds (or 175 gallons) of water.[16] Unless something is done to lower these consumption rates, modern agriculture will help to propel the United States into a water crisis.

Overall, it takes 1,000 tons of water to grow 1 ton of grain.[17] This is an average, with rice being the thirstiest crop and corn the least thirsty. Nearly two out of every five tons of grain produced worldwide go to meat and poultry production.[18] This is just one reason why a non-meat diet could feed twice as many people as a diet including meat. We could save ourselves a lot of suffering by simply eating lower on the food chain.

Urbanization will increasingly come into competition with agriculture for water supplies, as the urban population continues to expand in the coming years, in large part at the expense of rural population. Many studies expect urban water use to double during the next 25 years.[19] Water supplies in the western United States already are being diverted from agriculture to thirsty cities willing to pay a premium for water. Tucson, Phoenix, and other Arizona cities have purchased water rights from more than 232,000 hectares of farmland. In Pima County, where Tucson is located, irrigation is expected to disappear entirely by 2020.[20]

Finally, agricultural runoff is one of the most significant sources of water pollution. Artificial fertilizers, particularly nitrogen fertilizers, lead to increased algae production in lakes, rivers, seas, and estuaries. The resulting population explosion of algae and other microorganisms leads to oxygen depletion, resulting in dead zones where fish, shrimp, and other creatures cannot survive. Dead zones are spreading offshore from many of the world's great river deltas, such as the Mississippi.[21]

Pesticide runoff is also a major source of water pollution. The US Environmental Protection Agency has found 98 different pesticides, including DDT, in groundwater in 40 states, contaminating the drinking water of over 10 million residents.[22]

4

Eating Fossil Fuels

SOLAR ENERGY is a renewable resource limited only by the inflow rate from the sun to the earth. Fossil fuels, on the other hand, are a stock-type resource that can be exploited at a nearly limitless rate. However, on a human time scale fossil fuels are nonrenewable. They represent a planetary energy deposit that we can draw from at any rate we wish, but which will eventually be exhausted without hope of renewal. The Green Revolution tapped into this energy deposit and used it to increase agricultural production.

Total fossil fuel use in the United States has increased twentyfold in the last four decades. In the US, we consume 20 to 30 times more fossil fuel energy per capita than people in developing nations. Agriculture directly accounts for 17 percent of all the energy used in this country.[1] As of 1990, we were using approximately 1,000 liters of oil to produce food from one hectare of land.[2]

In 1994 David Pimentel and Mario Giampietro estimated the output/input ratio of agriculture to be around 1:4.[3] For 0.7 kilogram-calories (kcal) of fossil energy consumed, US agriculture produced 1 kilocalorie of food. This estimate results in net food energy production of only 30 percent. The input figure for this ratio was based on the United Nations' Food and Agriculture Organization (FAO) statistics, which consider only fertilizers (without including fertilizer feedstock), irrigation, pesticides (without including pesticide feedstock), and machinery and fuel for field operations. Other agricultural energy inputs not considered were energy and machinery for drying crops, transportation for inputs and outputs to and from the farm, electricity, and construction and maintenance of farm buildings and infrastructures. Adding in estimates for these energy costs brought the output/ input energy ratio down to 1:1.[4] Yet this does not include the energy expense of packaging, delivery to retail outlets, refrigeration, or household cooking.

In a subsequent study completed later that same year, Giampietro and Pimentel managed to derive a more accurate ratio of the net fossil fuel energy ratio of agriculture.[5] In this study, the authors defined two separate forms of energy input: endosomatic energy and exosomatic energy. Endosomatic energy is generated through the metabolic transformation of food energy into muscle energy in the human body. Exosomatic energy is generated by transforming energy outside of the human body, such as burning gasoline in a tractor. This assessment allowed the authors to look at fossil fuel input alone and in ratio to other inputs.

Before the industrial revolution, virtually 100 percent of both endosomatic and exosomatic energy was solar driven. Fossil fuels now represent 90 percent of the exosomatic energy used in the United States and other developed countries.[6] The typical exo/ endo ratio of pre-industrial, solar powered societies is about 4 to 1. The ratio has changed tenfold in developed countries, climbing to 40 to 1. And in the United States it is more than 90 to 1.[7] And the nature of the way we use endosomatic energy has changed as well.

The vast majority of endosomatic energy is no longer expended to deliver power for direct economic processes. Now the majority of endosomatic energy is utilized to generate the flow of information directing the flow of exosomatic energy driving machines. Considering the 90:1 exo/endo ratio in the United States, each endosomatic kcal of energy expended in the US induces the circulation of 90 kilocalories of exosomatic energy. As an example, a small gasoline engine can convert the 38,000 kilocalories in one gallon of gasoline into 8.8 Kilowatt hours, which equates to about 3 weeks of work for one human being.[8]

In their refined study, Giampietro and Pimentel found that ten kilocalories of exosomatic energy are required to produce one kilocalorie of food delivered to the consumer in the US food system. This includes packaging and all delivery expenses, but excludes household cooking.[9] According to this study, the US food system consumes ten times more energy than it produces. Giampietro and Pimentel's study is in agreement with an earlier study by C. A. S. Hall, C. J. Cleveland, and R. Kaufman.[10] However, Martin C. Heller and Gregory A. Keoleian arrive at a somewhat lower figure of 7.3 units of fossil fuel energy to produce one unit of food energy.[11] Even this latter figure represents a tremendous energy deficit.

Heller and Keoleian also looked at the energy cost of restaurants and household storage and preparation. They found that in 1995, commercial food services consumed 332 trillion British thermal units; nearly a third of this amount was used in cooking, the remainder went to refrigeration and various other uses. They also found that household storage and preparation consumed more energy than any other sector of the food system. The household energy demand was nearly 48 percent more than the energy required for agricultural production. Over 40 percent of household food-related energy consumption is used in refrigeration, 20 percent is used in cooking, and another 20 percent in heating water (largely for dishwashing).[12]

A 1976 study tracing the energy costs of moving corn through our food system found that more energy is used to drive to the supermarket to buy a can of corn than is consumed in producing the corn (assuming the crop was not irrigated). This study also found that if corn is kept frozen for more than 22 days, more energy is used than if it were canned.[13]

Giampietro and Pimental's 10:1 ratio means that it takes only 20 minutes of labor to provide most Americans with their daily diet — as long as that labor is fossil fueled. Unfortunately, if you remove fossil fuels from the equation, the daily diet will require 111 hours of endosomatic labor per capita; that is, the current US daily diet would demand nearly three weeks of work from each American to produce the amount of food they eat each day.[14]

Quite plainly, as fossil fuel production begins to decline within the next decade, there will be less energy available for the production of food.

US Consumption

In the United States, each person consumes an average of 2,175 pounds of food per year. This provides the US consumer with an average daily energy intake of 3,600 calories. The world average is 2,700 calories per day.[15] Every day, a quarter of the US population eats fast food (based on survey data gathered from 1994–1996). This figure is up from one-sixth of the population reported in a similar survey five years earlier. Fast food provides up to one-third of the average US citizen's daily caloric intake.[16] And there is good reason to believe that these figures have gone up since the years when this survey was conducted.

One third of the caloric intake of the average American comes from animal sources (including dairy products), totaling 800 pounds per person per year. This diet means that US citizens derive 40 percent of their calories from fat — nearly half of their diet.[17]

Americans are also grand consumers of water. As of one decade ago, Americans were consuming 1,450 gallons/day/capita (g/d/c), with the largest amount expended on agriculture. Allowing for projected population increase, consumption by 2050 is expected to drop to 700 g/d/c, which hydrologists consider minimal for human needs.[18] This is without taking into consideration declining fossil fuel production. As it declines, less energy will be available to power irrigation equipment.

To provide all of this food requires the application of 1.2 billion pounds of pesticides in the United States per year. The US accounts for one-fifth of the total annual world pesticide use, estimated at between five and six billion pounds. This equates to five pounds of pesticides for every man, woman and child in the nation.[19] Among the hydrocarbon-based pesticides are Methyl Parathion, Aldrin, Dieldrin, Endrin, Endosulfan, Chlordane, DDT, Heptaclor, Kepone, Lindane, Mirex, and Toxaphene. There are many others. All of them are powerful neurotoxins and are very persistent in the environment, due to their complex hydrocarbon backbones. Monsanto's Roundup, the most widely used herbicide in the world, is not hydrocarbon-based. It is, rather, a chain built on phosphorus and nitrogen. However, Monsanto's claims that Roundup is nontoxic for humans are being disproved. Roundup has been linked to non-Hodgkin's lymphoma.[20]

According to a recent study based on data supplied by the Centers for Disease Control, virtually every resident of the US has pesticide residue in their body. The average person has 13 pesticides or pesticide breakdown products in their body. The most prevalent chemicals — found in virtually all of the test subjects — are TCP (a metabolite of the insecticide chlorpyrifos, commonly known by the product name Dursban) and p,p-DDE (a breakdown product of DDT). Children contained the most pesticides, followed by women and Mexican-Americans. Many of these pesticides are present in amounts well in excess of established safety levels.[21]

In the last two decades, the use of hydrocarbon-based pesticides in the US has increased thirty-three-fold, yet each year we lose more crops to pests.[22] This is the result of abandoning traditional crop rotation practices. Nearly 50 percent of US corn is grown as a monoculture.[23] This results in an increase in corn pests that in turn requires the use of more pesticides. Pesticide use on corn crops had increased a thousandfold even before the introduction of genetically engineered, pesticide-resistant corn. However, corn losses have still risen fourfold.[24]

Worldwide, more nitrogen fertilizer is used per year than can be supplied through natural sources. Likewise, water is pumped out of underground aquifers at a much higher rate than it is recharged. And stocks of important minerals, such as phosphorus and potassium, are quickly approaching exhaustion.[25]

Total US energy consumption is more than three times the amount of solar energy harvested as crop and forest products. The United States consumes 40 percent more energy annually than the total amount of solar energy captured yearly by all US plant biomass. Per capita use of fossil energy in North America is five times the world average.[26]

Our prosperity is built on the principal of exhausting the world's resources as quickly as possible, without any thought to our neighbors, other life forms on this planet, or our children.

Food Miles

Food miles represent the distance food travels from where it is produced to where it is consumed. Food miles have increased dramatically in the last couple of decades, largely as a result of globalization. In 1981, food journeying across the US to the Chicago market traveled an average of 1,245 miles; by 1998, this had increased 22 percent, to 1,518 miles.[27] In 1965, 787,000 combination trucks were registered in the United States, and these vehicles consumed 6.658 billion gallons of fuel. In 1997, there were 1,790,000 combination trucks that used 20.294 billion gallons of fuel.[28] In 1979, David and Marcia Pimentel estimated that 60 percent of all food and related products in the US traveled by truck and the other 40 percent by rail.[29] By 1996, almost 93 percent of fresh produce was moved by truck.[30]

These studies only consider food traveling inside the United States. When we take into consideration all the food imports, and the distance they travel to reach their destination, the figure for food miles grows prodigiously. In the three decades from 1968 to 1998, world food production increased 84 percent, world population increased 91 percent, but food trade increased 184 percent.[31] An increasing percentage of the food eaten in the US is grown in other countries, including an estimated 39 percent of fruits, 12 percent of vegetables, 40 percent of lamb, and 78 percent of fish and shellfish in 2001.[32] The typical American prepared meal contains, on average, ingredients from at least five other countries.[33] Overall, agricultural imports into the US increased 26 percent by weight from 1995 to 1999.[34]

Using a measure of weighted average source distance (WASD), one study found that produce destined for consumers in Toronto, Canada traveled an average of 3,333 miles.[35] Computing food miles is not as straightforward as it would seem. For instance, most of the food coming into the US and Canada travels through

Los Angeles. Even food distributed within North America is first shipped to LA. So pears and apples from Washington State, right next to the Canadian border, make a longer journey to reach Toronto than carrots from California.[36]

A study of table grapes found that in 1972–73 the WASD of table grapes bound for Iowa was 1,590 miles. By 1988–89, this distance had increased to 2,848 miles. This is mostly explained by an increase in exports of Chilean table grapes. In 1998–99, the distance had diminished slightly to 2,839 miles, due to an increased percentage of Mexican table grape imports.[37]

This phenomenon is not confined to North America. In the UK, the distance traveled by food increased 50 percent between 1978 and 1999.[38] A Swedish study of the food miles involved in a typical breakfast (apple, bread, butter, cheese, coffee, cream, orange juice, sugar) found that the mileage estimated for the entire meal was equivalent to the circumference of the Earth.[39]

The increase in food miles is, of course, made possible by an increase in fossil fuel consumption. So the globalization of food production and the atrophying of localized food infrastructure are subsidized by cheap and abundant fossil fuels. As fossil fuels become less abundant and more expensive, this system will become increasingly strained until it finally collapses, leaving local communities without the ability to feed themselves.

The globalization of food has an adverse effect on local farmers as well. From the indigenous farmers of third world countries who can no longer compete with cheap grain imports from the US, to the farms of the American Midwest that are losing their agricultural diversity, food security is threatened by globalization. Consider Iowa, which is blessed with some of the richest farmland the world over.

In 1920, Iowa produced 34 different commodities on at least 1 percent of its farms, and 10 different commodities were produced on over half of its farms.[40] In 1870, virtually all the apples consumed in Iowa were grown locally.[41] By 1999, only 15 percent of the apples consumed in Iowa were produced locally.[42] Overall, by the 1970s, one percent of Iowa farms were producing no fruit or vegetables. By 1997, only corn and soybeans were produced on over half of Iowa's farms.[43] By 1998, the state that had once been the world leader in canned sweet corn production had only two remaining canning facilities in the entire state.[44] Excepting meat production, most Iowa farms no longer produce food to supply Iowa consumers directly.[45]

Globalized food production is a threat to food security, not only because of the collapse of local food production, but through importing diseases, invasive species, and poisons. The latter is due to the overseas use of pesticides that have been banned for use within the United States. The increasing distance between food production and markets also promotes consumer ignorance, as consumers are isolated from the true social and environmental costs of the products they consume.

The only way to grasp all these unacknowledged expenses is through a life-cycle assessment, which is a very complicated procedure. As an example, the life-cycle assessment of coffee production shows that a morning cup of Colombian java has the following global effects:

- In the Antioquia region of Colombia endangered cloud forests are cleared and the watershed polluted several times annually by the application of pesticides.
- In Europe's Rhone River Valley the effluents of pesticide production have helped to turn the Rhone River into one of the most heavily polluted rivers in the world.
- In Papua New Guinea, iron is mined from stolen tribal lands for the ships that transport the beans, leaving the land disturbed and polluted.
- In New Orleans the beans are roasted and packaged in plastics made from oil shipped by tanker from Venezuela and the Middle East and produced at factories in Louisiana's "cancer corridor," with its disproportionately black surrounding population.
- In the ancestral lands of the Australian Aborigines bauxite ore is strip-mined for the packaging's aluminum layer.
- On the west coast of the USA the bauxite is refined using hydroelectric power from the Colombia River, the harnessing of which destroyed local American Indian salmon fisheries.

- Near Philadelphia oil to transport the packaged coffee is refined at a plant where heavy air and water pollution have been linked to cancer clusters, contaminated fish, and a decline of marine wildlife throughout the Delaware River basin.
- All over the continent that cup of coffee depends on oil, natural gas, and coal in a hundred other incidental ways — including lighting, heating, and cooling the high-rise offices of advertising and food company executives, as well as the media executives whose magazines and TV shows carried ads for the coffee.[46]

Similar life-cycle assessments can be made for any globalized food commodity. All of them will show environmental destruction and degradation, the exploitation of indigenous peoples and cultural minorities, and the wasteful consumption of fossil fuels. This system is destructive and unsustainable.

Not only is our entire agricultural and food system based upon the availability of cheap fossil fuels — we do not even use them in a wise and frugal manner. We squander them on flagrant consumerism in order to maximize short-term profit, while destroying the localized systems that once sustained our culture.

5

The End of the Oil Age

CURRENT CIVILIZATION is founded upon an abundance of cheap energy derived from hydrocarbons. Hydrocarbons not only run our transportation; they provide the power for all of our technology. Take a moment to think about your immediate home environment. Not only do hydrocarbons take you to work and to the grocery store; they are used for virtually everything around you. Your home and your furniture were built using the energy of hydrocarbons. If your chair has a metal frame, that metal was forged with hydrocarbons. Your carpet and your polyester clothing are products of hydrocarbons. All of the plastics around you are derived from hydrocarbons. Even this book was printed and delivered using hydrocarbons. The very value of the money in your wallet is pegged to oil. We have seen in previous chapters how dependent modern agriculture is upon oil and natural gas.

Before the industrial revolution, all civilizations were built on the energy of slave labor. Even the United States required the sweat of slaves during its early years. The industrial revolution rendered slavery, and all other forms of servitude, obsolete. First it was coal that supplied the power to run our furnaces. But eventually coal was replaced by oil, with its far superior caloric content. It was the late 1800s when we began to seriously exploit oil resources. This abundance of cheap, high-content energy gave rise to the technological revolution of the last hundred years. There are various estimates of the "slave equivalents" of cheap hydrocarbon-based energy, but there is no doubt that every one of us is served by a multitude of hydrocarbon slaves.

Earlier in this book, we discussed how oil was formed by a combination of biological and geological processes, dependent upon special environmental circumstances that no longer exist. Through this bio-geological process, solar energy was stored and converted into hydrocarbons over millions of years.

Now enter the humans. Here is an exploration team fortunate enough to discover a sizable oil reserve. Wells are drilled to tap into the oil. Now oil in the ground flows at about the same rate as damp in a stone foundation, the one major difference being that the oil is held at much higher pressure. When a new well is drilled, the open hole gives free passage to the pressurized oil immediately around it, which then surges to the surface. The effect is the classic gusher featured in so many films. However, once the pressure has been equalized in the immediate vicinity of the drill hole, it takes more and more energy to pump the oil through the rock or sediment to the well. Eventually you will reach a point where you must invest as much energy to pump the oil as you will get out of it. When this point is reached, production ends and the well is capped forever.

If you draw a graph of oil production over time, it will resemble a classic bell curve. The production curve will start from nothing, ascend to a peak production rate, and then begin to descend. The descending side of the curve means that you are investing more energy to produce the oil, which makes the oil more expensive. During the 1950s and 1960s a petroleum geologist named M. King Hubbert developed a methodology for combining the profiles of oil wells in a field to draw a production curve for the entire field. From there, he went on to develop production curves for regions and even countries. Using industry data, Dr. Hubbert was able to tie his production curves to the discovery rate with a lag time of about thirty years.

Using this methodology, Dr. Hubbert predicted that oil production in the United States would peak in 1970. M. King Hubbert was ridiculed and condemned for his prediction. The conventional wisdom, as espoused by the US Geological Survey (USGS), was that oil production would continue to rise for many years to come. Unfortunately for us, Hubbert was correct. Oil production in the United States peaked in the early 1970s and has been declining ever since. Right now we are importing over half of our oil needs. The US production peak of the early 1970s set the stage for the oil shocks of that decade and the rise of OPEC. However, at that time we were able to increase imports to make up the difference between domestic production and demand. Alaskan and North Sea oil were brought online soon enough to defang OPEC, and the US became the oil protector of the world by forcing OPEC to accept the dollar as the currency for oil sales.

A number of predictions were made over the years following the US peak regarding the global oil production peak. Many people now look back on these false predictions and use them to condemn the current scientific consensus. However, none of these early predictions were actually made by oil geologists with access to the database of the oil industry. Over the years the methodology has been improved and the database has been augmented.

In the 1990s, oil geologists finally felt confident enough in the data to draw up graphs for world oil production. Two leaders in this effort are Colin J. Campbell and Jean H. Laherrère, petroleum geologists working for Petroconsultants. Petroconsultants holds one of the most complete databases in the industry. In 1997, Petroconsultants' annual report on the state of the oil industry (which costs a whopping $10,000 per copy) dealt strictly with the topic of peak production and predicted that world production would peak in the first decade of the new century and begin its irreversible decline sometime around 2010.

There have been several other independent assessments since then, and most agree on the timeframe. The most notable dissenting voice is the 2000 USGS report which stated that world oil production would not peak until 2020 at the earliest. However, it has been shown that the USGS study is deeply flawed.[1] It accepted as valid oil reserves any reserve with a ten percent or greater chance of being discovered. The realistic benchmark is 50 percent. In the few years since publication, the USGS report has already proven unreliable in comparison with actual production and discovery rates.

There are a lot of problems with oil production data. The industry has a tendency to underreport initial discoveries so that they can add them on later to give the impression of a steady discovery rate. A steady discovery rate looks more appealing to investors. OPEC countries, on the other hand, have a marked tendency to inflate their oil reserves when it comes time to adjust quotas under OPEC. And politically, nobody wants to let the general population know that the party is almost over. The US Energy Information Administration (EIA) has publicly stated — although in a roundabout manner — that they first project future energy demand and then they come up with reserve and production figures to meet their projected demand.[2] Making sense of the data requires a lot of detective work, but a scientific consensus has been achieved.

This is the standard espoused by Campbell and Laherrère. According to their scenario, we are at peak production right now. Currently, we are engaged in a tango between world oil production and the global economy. Rising oil prices lead to economic stagnation and a decrease in demand, which then leads to lower production and a softening in oil prices, until economic rebound results in demand once again rising above production. Of course, this is a simplified model. It would take much more space to add in all the other economic and oil-related factors, not to mention the effects of oil wars.

However, the major oil companies have started making coded announcements indicating that they know the future of the oil business will not match its past. Instead of investing in production and discovery, all of the majors have been shedding exploration staff and consolidating their holdings. None of this bespeaks a growing industry. And insiders know that there is very little excess capacity to be found anywhere.

There was considerable hope prior to the Afghan War that the Caspian Sea held oil reserves that would match — if not dwarf — the Middle East. However, exploration has produced disappointing results. The Caspian region does not hold nearly as much oil

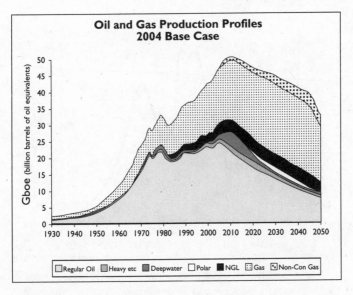

as was at first supposed, and the oil that has been found is highly tainted by sulfur. As a result, the oil majors have been scaling back their involvement in the Caspian region.

Some point to Russia as a rival to Saudi Arabia. This ignores the reality that Russia is simply resuming a production capacity that faltered following the collapse of the Soviet Union. Russian production peaked in 1987.[3] This is illustrated in the following graph, showing the peak of Russian oil production in the late 1980s, followed by a sharp drop in large part due to the collapse of the former Soviet Union. This is then followed by a sharp rise in production through the 1990s and up to the present time. Production has risen so sharply because of the revival of the Russian oil industry and the implementation of aggressive production techniques. This aggressive Russian production will be paid for by a quicker peak and a steeper decline. *The Moscow News* has reported that Yuri Shafranik, the head of the Russian Union of Oil and Gas Producers, stated on November 9, 2004, that Russia has almost reached its maximum production and the decline will start within two years. Mr. Shafranik referred to experts from the International Energy Agency.[4]

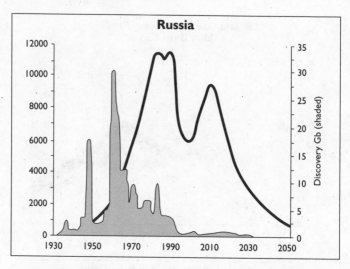

Richard Duncan and Walter Youngquist developed a system in the late 1990s to model world oil production. As they ran simulations with this model, they attempted adding on additional units of oil, each unit equivalent to a reserve the size of the North Sea. Additional units brought into production after the peak had no effect on peak production. But they found that several additional units brought into production before the peak could delay the peak by a year or two. This would be the equivalent of several new North Sea discoveries.[5] Yet oil exploration geologists warn that all we will find from now on are small isolated pockets. All that our knowledge and technological advancement have managed to show us is where oil does not exist.

Still, there are economists who will tell you that it is only a matter of money. If we throw enough money into exploration and development, we will increase production. This seems to belie actual experience. Over the last thirty years increased investment and technological advances have led to only marginal gains in discovery and production. Were it otherwise, the industries would not be scaling back.

Others say that we will abandon hydrocarbons for better energy sources. This ignores the fact that there is no other energy resource capable of delivering as much energy as hydrocarbons

— not renewables, not unconventional resources such as tar sands, not even coal. The only thing which comes close is nuclear, and this has too many other problems.

There are many true believers — including former Secretary of Energy Spencer Abraham — who point to a world run by fuel cells. What fuel cell proponents won't tell you is that hydrogen fuel cells are not an energy *source*. They are more properly a form of energy *storage*. In the natural world there is no such thing as free hydrogen. Hydrogen must be produced from a feeder material. Nor is it mentioned that it takes more energy to break a hydrogen bond than can be gained through the forging of a hydrogen bond. This is basic chemistry, as implied in the Second Law of Thermodynamics. As a result, hydrogen fuel cells will always have a net energy loss. Nor are they as clean as claimed. The pollution is simply removed from the individual vehicles to the plant where free hydrogen is generated. It is most likely that the hydrogen fuel cell myth is being promoted simply to keep the public — and investors — from panicking.

The truth is that peak oil has already had an impact on all of the major events of this young century. And it will have a major impact on all of our lives at a most personal level in the years to come. This impact will be felt not only at the gas pumps. The impact of fossil fuel depletion will be felt in higher food prices and, if nothing is done to completely revamp our food system, sooner or later in food shortages and massive starvation. The public needs to be informed. Our civilization is about to undergo a radical change unparalleled in history. And those we are allowing to call the shots are more concerned with their own personal gain than with the general welfare.

The Natural Gas Cliff

Natural gas is every bit as important for agriculture as is oil, perhaps even more so. All artificial fertilizers are derived from natural gas. What we call natural gas is actually a mixture of hydrocarbon gases. Methane is one of the main components. Fertilizer production converts methane to ammonia, which is then used to produce ammonia-based fertilizers. Natural gas is also used to power some irrigation systems, and for indoor heating of meat factories.

The situation for natural gas production differs from the situation for oil, but it is not any brighter. While pumping oil from the ground is somewhat like straining molasses through a sand sieve, pumping natural gas is more like poking a hole in a car tire. This is due to the fact that oil is a liquid and natural gas is a gas. When you poke a hole in a car tire the air will escape for some time, flowing freely outward until pressure is equalized between the outside atmosphere and the interior of the tire. To extract the remaining air from the tire, you must attach it to a vacuum pump, or squeeze the inner tube.

Tapping into a body of natural gas is generally less costly than tapping into an oil field. Once the wells are drilled, you simply have to hook them to a pipeline. Natural gas production increases rapidly from the time a field is first put into production, rising until the field is fully covered with producing wells. Then production flattens out and continues at that level for an unpredictable length of time. Once the production of a field has flattened out, it is difficult to increase it further. If you wish to increase production, you must find another field. At some unknown point, production in the field will fall into a marked decline. The decline rate for natural gas fields is much higher than the decline rate for oil fields; somewhere in the neighborhood of five to ten percent, compared to oil's two or three percent. Because the decline is so steep, it is known as the natural gas cliff. There is little warning of the cliff in the field's production. The last square foot of gas to be extracted from a field before production falls off the cliff will require no more effort than the first square foot extracted from it.

Because we are dealing with a gas here, measurements are different than those we use for oil. Oil is measured in barrels or metric tonnes. Natural gas is measured in cubic feet, most commonly in billions of cubic feet (Bcf), or trillions of cubic feet (Tcf). A trillion cubic feet may sound like a lot, but we must remember that gas is less dense than oil so it holds less energy. US demand for natural gas is expected to rise to 30 Tcf per year by

2010. At one time, it was believed that most of this would come from the Gulf of Mexico, but the US Minerals Management Service now expects Gulf production to begin declining in 2005, from a peak of 6.1 Tcf per year, by at least 5 to 7 percent.[6]

Overall, the North American outlook for natural gas production is not good. Mexican production flattened out in 2002. Mexico stopped exporting natural gas to the US in 2000, and has been a net importer of natural gas ever since.[7] US production has been at a plateau for some time. All the big finds have been tapped and are in decline. US production history shows that new wells are being depleted more quickly all the time; the current decline rate is 28 percent. While this is partially due to growing demand, it is also due to the fact that the large deposits of natural gas are all aging and are in terminal decline. Newer deposits tend to be smaller and are produced (and depleted) quickly in the effort to maintain overall production levels.[8]

The United States turns to Canada to make up the difference between its own flattened production and rising demand. Canada currently supplies at least 13 percent of the US gas demand. Yet Canada's large fields have also flattened out in production, and it is likely that Canadian production will fall off the cliff within the next several years.[9]

Worldwide, natural gas production will not begin to decline for at least another decade, and by some estimates not for 20 to 30 years. However, because we are talking about a gas, world production is not as important as regional production. We must look to North American natural gas production to meet the lion's share of our needs. Natural gas is most easily transported in pipelines; it is very difficult to transport overseas. The only effective way to ship it is to liquefy it, transport it in specially designed refrigerated tankers, and then unload it at specially designed facilities that will thaw it back to the gaseous state. All of this is done at an estimated 15 to 30 percent energy loss.

Currently, there are only four liquid natural gas (LNG) offloading facilities in the US, located in Louisiana, Georgia, Maryland, and Massachusetts. In 2003, we imported an average of 1.5 Bcf per day (Bcf/day). This amounted to 2 percent of our natural gas demand of 67 Bcf/day. By the end of 2006, we are hoping to add another three Bcf/day of LNG imports. But by the end of the decade, demand is expected to rise to 77 Bcf/day.[10]

Today the global fleet of LNG tankers numbers 140, with a capacity of 14.5 Bcf/day. By the end of the decade, the US will require this entire fleet just to service our needs. LNG tankers cost an average of $155 million per ship to build. So the tanker fleet alone will require an investment of $13 billion. Add to this the expense of building over 30 new LNG projects and the associated pipelines, and the necessary investment quickly climbs over $100 billion. Considering our current budget deficit and the precarious state of the US economy, on top of the fact that world natural gas production will peak in another 10 to 30 years, this sort of investment is unlikely.[11]

This is why politicians and the corporations who pay them are clamoring to open currently restricted areas of Alaska, the Canadian Arctic, the US Rocky Mountains, and the deep ocean to natural gas development. Yet the eventual investment in pipelines and drilling rigs to tap these sources would be even higher than the cost of LNG development: an estimated $120 billion in infrastructure. And from the time construction begins on this infrastructure, it will take five to seven years before any of this gas begins to flow. In total, we are talking about less than a decade's worth of natural gas here, even at our current rate of demand.[12]

As natural gas becomes more expensive and harder to acquire, we must find some substitute to serve as fertilizer. If substitutes cannot be provided in the same proportion, then we cannot expect to grow enough crops in our depleted soils. It just so happens that there is an abundant, natural fertilizer that is currently going to waste: manure. Later in this book, we will talk about the necessity of closing the nutrient cycle by recycling animal and human manure. But this effort will require a major investment in infrastructure — particularly in processing facilities to compost the manure and purge it of harmful pathogens and pollutants.

The North American natural gas cliff is the other side of the approaching energy crisis. Our economy, and our very lifestyle, is caught between the gas cliff and the oil peak. Between them, they are going to make life very difficult in the years to come.

6

The Collapse of Agriculture

Modern industrial agriculture is unsustainable. It has been pushed to the limit and is in danger of collapse. As we saw earlier in this book, we have already appropriated all of the prime agricultural land on this planet; all that remains is a small percentage of marginal lands and those areas — deserts, mountains, polar regions — that are completely unsuitable. As a result, biological diversity — the underpinning of life on this planet — has been diminished nearly to the breaking point.

Moreover, our soils and fresh water resources have been degraded and depleted nearly to the crisis point. Our farm crops have been genetically reduced to weak, high-yield hybrids that are susceptible to any number of pests, and that offer a minimum of nourishment. Our land and water resources, and even our food, are also highly tainted with toxins we have over-applied in an effort to protect our food crops from pests. And our farmlands have been concentrated into agribusinesses dedicated to maximizing short-term profit — while, incidentally, undermining our ability to support ourselves with local agriculture.

Even without considering energy depletion, our agricultural system is ready to collapse. Yet, the abundance of cheap food given to us by the Green Revolution has resulted in an exponential population boom. So we must now address a very serious question. Without the cheap, abundant supply of fossil fuels that has made possible the industrialization of agriculture, and that has allowed an explosion in food production at an energy deficit of ten to one, has the human population exceeded the carrying capacity of the planet? And if so, by how much?

Population and Sustainability

Assuming a growth rate of 1.1 percent per year, US population is projected to double by 2050. As the population expands, an estimated one acre of land will be lost for every additional person. Currently, 1.8 acres of farmland are available to grow food for each US citizen. By 2050, this will decrease to 0.6 acres. However, 1.2 acres per person is required to maintain current nutritional standards.[1]

Presently, only two nations on the planet are major exporters of grain: the United States and Canada.[2] By 2025, it is expected that the US will cease to be a food exporter due to domestic demand. The impact on the US economy could be devastating, as food exports earn $40 billion annually. More importantly, millions of people around the world could starve to death without US food exports.[3]

In the US, 34.6 million people were living in poverty according to 2002 census data.[4] This number continues to grow at an alarming rate. Too many of these people do not have enough food. As the situation worsens, this number will increase and the United States could witness growing numbers of starvation fatalities.

There are some things that we can do to at least alleviate this tragedy. It's been suggested that streamlining agriculture to get rid of losses, waste, and mismanagement might cut the energy inputs for food production by up to one-half.[5] In place of fossil fuel-based fertilizers, we could use livestock manures that are now being wasted. It is estimated that livestock manures contain five times the amount of nutrients fertilizers currently provide each year.[6] Perhaps the most effective step would be to eliminate meat from our diet altogether.[7]

Mario Giampietro and David Pimentel postulate that a sustainable food system is possible only if four conditions are met.

1. Environmentally sound agricultural technologies must be implemented.
2. Renewable energy technologies must be put into place.
3. Major increases in energy efficiency must reduce exosomatic energy consumption per capita.
4. Population size and consumption must be compatible with maintaining the stability of environmental processes.[8]

Providing that the first three conditions are met, with a reduction to less than half of the exosomatic energy consumption per capita, the authors place the maximum US population for a sustainable national economy at 200 million.[9] Several other studies have produced figures within this ballpark.[10] Given that the current US population is more than 297 million,[11] that would mean a reduction of 97 million. To achieve a sustainable economy and avert disaster, the United States must reduce its population by at least one-third. The black plague during the 14th century claimed approximately one-third of the European population (and more than half of the Asian and Indian populations), plunging that continent into a darkness from which it took them nearly two centuries to emerge.[12]

Personally, I can only hope that a fairer distribution of wealth and resources will help ease us down the path of depopulation. And I hope that the decline rate will be gradual enough that our population can shrink with the least amount of suffering. But I can only hope, recognizing sadly that depopulation runs contrary to the basic drive to procreate, and that depopulation has never been managed before without a die-off.

None of this research considers the impact of declining fossil fuel production. At the time of these studies, the authors believed that the agricultural crisis would only begin to impact us after 2020, and would not become critical until 2050. The current peaking of global oil production (and subsequent decline of production after 2010), along with the peaking of North American natural gas production, will very likely precipitate this agricultural crisis much sooner than expected. Quite possibly, a US population reduction of one-third will not be effective for

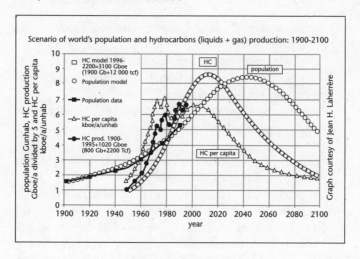

Scenario of world's population and hydrocarbons (liquids + gas) production: 1900-2100

sustainability; the necessary reduction might be in excess of one-half. And for sustainability, global population will have to be reduced from the current 6.5 billion people[13] to 2 billion — a reduction of 68 percent or over two-thirds.[14] The end of this decade could see spiraling food prices without relief. And the coming decade could see massive starvation on a global level such as never experienced before by the human race.

The Example of North Korea

What happens to an industrialized country when it loses its hydrocarbon base? Unfortunately, this very thing happened in North Korea. The Korean Peninsula has virtually no oil and no natural gas. North Korea relied on the Soviet Union for much of its energy needs. Following the crash of the Soviet Union, North Korea experienced a sharp and swift drop in its hydrocarbon imports. The effect was disastrous.

North Korea has always had less than half the population of South Korea. When the Korean peninsula was partitioned in 1945 at the end of World War II, creating North and South Korea, South Korea was a largely agrarian society, while the Democratic People's Republic of Korea (DPRK) was largely an industrial society. Following the war, DPRK turned to fossil fuel-subsidized agriculture to increase the productivity of its poor soils.

By 1990, DPRK's estimated per capita energy use was 71 gigajoules per person,[15] the equivalent of 12.3 barrels of crude oil. This was more than twice China's per capita usage at that same time, and half of Japan's. DPRK has coal reserves estimated at from one to ten billion tons, and developable hydroelectric potential estimated at 10–14 gigawatts.[16] But North Korea must depend on imports for all of its oil and natural gas. In 1990, it imported 18.3 million barrels of oil from Russia, China, and Iran.[17]

ENERGY CRISIS IN THE DPRK

Following the collapse of the Soviet Union, imports from Russia fell by 90 percent. By 1996, oil imports came to only 40 percent of the 1990 level.[18] DPRK tried to look to China for the bulk of its oil needs. However, China sought to distance itself economically from DPRK by announcing that all their commerce would be conducted in hard currency beginning in 1993. China also cut its shipments of "friendship grain" from 800,000 tons in 1993 to 300,000 tons in 1994.[19]

On top of the loss of oil and natural gas imports, DPRK suffered a series of natural disasters in the mid-1990s which acted to further debilitate an already crippled system. The years 1995 and 1996 saw severe flooding which washed away vital topsoil, destroyed infrastructure, damaged and silted hydroelectric dams, and flooded coal mine shafts rendering them unproductive. In 1997, this flooding was followed by severe drought and a massive tsunami. Lack of energy resources prevented the government from preparing for these disasters and hampered recovery.

DPRK also suffered from aging infrastructure. Much of its machinery and many of its industrial plants were ready for retirement by the 1990s. Because DPRK had defaulted on an enormous debt some years earlier, it had grave difficulty attracting the necessary foreign investment. The dissolution of the Soviet Union meant that DPRK could no longer obtain the spare parts and expertise to refurbish their infrastructure, leading to the failure of machinery, generators, turbines, transformers, and transmission lines. The country entered into a vicious positive feedback loop as failing infrastructure cut coal and hydroelectric production and diminished their ability to transport energy via power lines, truck and rail.

The decline in availability affected all sectors of commercial energy use between the years 1990 and 1996. As a result of this, North Koreans turned to burning biomass, thus destroying their remaining forests. Deforestation led, in turn, to more flooding and increasing levels of soil erosion. Likewise, soils were depleted as plant matter was burned for heat, rather than being mulched and composted.

By 1996, road and freight transport were reduced to 40 percent of their 1990 levels. Iron and steel production were reduced to 36 percent of 1990 levels, and cement was reduced to 32 percent.[20] The effect rippled out through the automotive, building, and agricultural industries. The energy shortage also affected residential and commercial lighting, heating, and cooking. This, in turn, led to loss of productivity and reduced quality of life, and adversely impacted public health. To this day, hospitals remain unheated in the winter, and lack electricity to run medical equipment. By 1996, total commercial energy consumption throughout society fell by 51 percent.[21]

Perhaps in no other sector was the crisis felt more acutely than in agriculture. The energy crisis quickly spawned a food crisis which proved to be fatal. Modern industrialized agriculture collapsed without fossil fuel inputs. It is estimated that over three million people have died as a result.[22]

THE COLLAPSE OF AGRICULTURE IN THE DPRK

The following graph, produced by Jean Laherrère, illustrates the relationship between petroleum consumption and agricultural collapse in DPRK.[23] Note that the decline of agricultural production follows very closely the decline of petroleum consumption. Also note that the rise in petroleum consumption after 1997 is not mirrored by a rise in agricultural production. Agriculture begins to make a comeback, but appears to enter another decline sometime around 1999. We do not have enough data at present to state

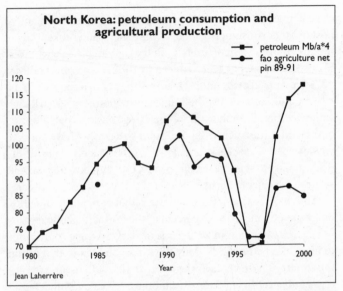

North Korea: petroleum consumption and agricultural production

- petroleum Mb/a*4
- fao agriculture net pin 89-91

Jean Laherrère

conclusively the reasons why this recovery has faltered. It is likely a combination of factors, such as failure of farm equipment and infrastructure, adverse weather, and — quite likely — the failure of soils which have been depleted of minerals over the past decade. In any case, the above graph sums up the agricultural collapse of DPRK and hints at the suffering that collapse has entailed.

Fertilizer

Agriculture in DPRK requires approximately 700,000 tons of fertilizer per year.[24] The country used to manufacture 80 to 90 percent of its own fertilizers, somewhere from 600,000 to 800,000 tons per year. Since 1995, it has had difficulty producing even 100,000 tons per year. Aid and foreign purchases brought the total for 1999 to 160,000 tons, less than one quarter of the required amount.[25]

The DPRK fertilizer industry relies on coal as both an energy source and a feedstock. It requires 1.5 to 2 million tons of coal per year to produce 700,000 tons of fertilizer.[26] To obtain this coal, the fertilizer industry must compete with the steel industry, electricity generation, home heating and cooking needs, and a host of other consumers. Flooded mine shafts and broken-down mining equipment have severely cut the coal supply. Likewise, delivery of this coal has been reduced by the breakdown of railway infrastructure. Furthermore, transporting two million tons of coal by rail requires five billion kilowatt hours of electricity,[27] while electricity is in short supply because of lack of coal, silting of dams, and infrastructure failure. So once again we have another vicious positive feedback loop. Finally, infrastructure failure limits the ability to ship the fertilizer — 1.5 to 2.5 million tons in bulk — from factories to farms.[28]

The result of this systemic failure is that agriculture in DPRK operates with only 20 to 30 percent of the normal soil nutrient inputs.[29] The reduction in fertilizer is the largest single contributor to its reduced crop yields. Tony Boys has pointed out that to run the country's fertilizer factories at capacity would require the energy equivalent of at least five million barrels of oil, which represents one quarter of all the oil imported into DPRK in recent years.[30] However, even capacity production at this point would be inadequate. For the past decade, soils in the DPRK have been depleted of nutrients to the point that it would now require a massive soil building and soil conservation program to reverse the damage.

Diesel Fuel

Agriculture has been further impacted by the limited availability of diesel fuel. Diesel fuel is required to run the fleet of approximately 70,000 tractors, 8,000 tractor crawlers, as well as 60,000 small motors used on small farms in DPRK.[31] Diesel is also required for transporting produce to market and for food processing equipment. It is estimated that in 1990, North Korean agriculture used 120,000 tons of diesel fuel. Since then, the amount used by agriculture has declined to between 25,000 and 35,000 tons per year.[32]

Compounding the diesel supply problem is its military allocation, which has not been cut proportionally with the drop in production. Only after the military takes its allocation can the other sectors of society — including agriculture, transportation, and industry — divide the remainder. So, while total supplies of diesel have dropped by 60 percent, the agricultural share of the remainder has fallen from 15 percent in 1990 to 10 percent currently.[33] In other words, agriculture must make due with 10 percent of 40 percent, or 4 percent of the total diesel supply of 1990.

The result is an 80 percent reduction in the use of farm equipment.[34] There is neither the fuel nor the spare parts to keep farm machinery running. Observers in 1998 reported seeing tractors and other farm equipment lying unused and unusable while farmers struggled to work their fields by hand. The observers also reported seeing piles of harvested grain left on the fields for weeks, leading to post-harvest crop losses.[35]

Lost mechanized power has been replaced by human labor and draft animals. In turn, due to their greater work load, human laborers and draft animals require more food, putting more strain on an already insufficient food supply. And although a greater percentage of the population is engaged in farm labor, they have found it impossible to perform all of the operations previously carried out by machinery.[36]

Irrigation

Finally, the agricultural system has also been impacted by the decreased availability of electricity to power water pumps for irrigation and drainage. The annual amount of electricity necessary for irrigation throughout the nation stands at around 1.2 billion kilowatt hours. Adding another 460 million kilowatt hours to operate threshing and milling machines and other farm equipment brings the total needed for agriculture up to 1.7 billion kilowatt hours per year.[37] This does not include the electrical demand for lighting in homes and barns, or any other rural residential uses.

Currently, electricity available for irrigation has declined by 300 million kilowatt hours, and electricity for other agricultural uses has declined by 110 million kilowatt hours. This brings the total electrical output currently available for agriculture down to 1.3 billion kilowatt hours — a shortfall of 400 million kilowatt hours from what is needed.

In reality, the situation for irrigation is even worse than indicated by these figures. Irrigation is time sensitive — especially in the case of rice, which is DPRK's major grain crop. Rice production depends on carefully timed flooding and draining. Rice is transplanted in May and harvested in late August and early September. From planting to harvest time, the rice paddies must be flooded and remain in water. In DPRK virtually all rice irrigation is managed with electrical pumps; over half of the pumping for all agriculture takes place in May. Peak pumping power demand at this time is at least 900 megawatts. This represents over one-third of the country's total generating capacity.[38] The energy crisis has had a severe impact on rice irrigation.

On top of this, the national power grid is fragmented, so that at some isolated points along the grid, irrigation demand can overtax generating capacity. This overtaxed system is also dilapidated, suffering the same disrepair as other energy infrastructure, both due to weather disasters, the age of the power stations and transmitters, and the lack of spare parts.

The records of three major pumping stations showed that they suffered an average of 600 power outages per year, spending an average of 2,300 hours per year without power. These power failures resulted in an enormous loss of water, translating into an irrigation shortfall of about one-quarter of the required amount of water.[39]

HOME ENERGY USAGE

Home energy usage is also severely impacted by the energy crisis, and — particularly in rural areas — home energy demand in turn impacts agriculture. Rural residential areas have experienced a 50 percent drop in electricity consumption, resulting in a decline in basic services and quality of life. Homes in rural villages rarely have electrical power during the winter months.[40] As has already been mentioned, hospitals and clinics are not excluded from this lack of power.

Rural households use coal for heating and cooking. The average rural household is estimated to require 2.6 tons of coal per year. The total rural coal requirement is 3.9 million tons annually. Currently, rural areas receive a little more than half of this.[41] On average, rural coal use for cooking, heating, and preparing animal feed has declined by 40 percent, down to 1.6 tons per year.[42] Even public buildings such as schools and hospitals have limited coal supplies.

To make up for the shortfall in coal, rural populations are increasingly turning to biomass for their heating and cooking energy needs. Herbage has been taken from competing uses such as animal fodder and compost, leading to further decreased food supplies. Biomass scavenging is also stressing all rural ecosystems from forests to croplands. Biomass harvesting reduces ground cover, disrupts habitats, and leads to increasing soil erosion and siltation.

Moreover, biomass foraging requires time and effort when other labor requirements are high and food supplies are low. This contributes to the positive feedback loop of calorie requirements versus food availability. It is estimated that 25 percent of the civilian workforce was employed in agriculture in the 1980s. By the mid-1990s, that had grown to 36 percent.[43] Furthermore, agricultural work has grown much more labor intensive. Farm labor is conservatively estimated at a minimum of 300 million person-hours per year. However, researchers point out that this number could easily be higher by a factor of two or more.[44] Workers are burning more calories and so require more food. This is further complicated by greater reliance upon draft animals with their own food requirements. So necessary caloric intake has actually increased as food production has decreased, leading to increasing malnutrition.

IMPACTS ON HEALTH AND SOCIETY

US congressmen and others who have visited North Korea tell stories of people eating grass and bark. Other reports talk of soldiers who are nothing more than skin and bones. Throughout the country there is starvation to rival the worst found in Africa. Chronic malnutrition has reached the point where many of its effects are irreversible.[45]

A study of children aged 6 months to 7 years found that 16 percent suffered from acute malnutrition — one of the highest rates of wasting in the world. Three percent of the children suffered edema, 61 percent were moderately or severely underweight and 62 percent suffered from chronic malnutrition, which can lead to irreversible stunting.[46]

Furthermore, malnutrition weakens the immune system, leaving the population even more vulnerable to contagions. And the lack of fuel for boiling water has led to a rise in water-borne diseases. Without electricity and coal, hospitals and clinics have become harbors of despair, where only the hopeless go for treatment.[47]

The situation in DPRK has rendered the country even more vulnerable to natural disasters. The country lacks the energy reserves to recover from the natural disasters of 1995-1997, much less to withstand future such disasters. The infrastructure is fragmented and in disrepair. There is a very real threat that portions of the infrastructure, such as the electrical grid, may fail altogether. Complete electrical grid failure would result in a near-complete crop loss.[48]

So far, the people of DPRK have faced this crisis together. But continued deprivation may very well lead to rivalry, regional fragmentation, social breakdown, and internecine fighting. Rural society is currently faring better than the urban population, and is actually absorbing urban workers to help meet the rising labor demands of agriculture. But worsening conditions and widespread flight from the cities could lead to violent confrontations. It is even possible that rural instability could eventually result in civil war.

A MODEL FOR DISASTER

The history of DPRK through the 1990s demonstrates how an energy crisis in an industrialized nation can lead to complete systemic breakdown. Only outside help has allowed DPRK to make any recovery. Of particular note is how the energy crisis sends ripples throughout the entire structure of society, and how various problems act to reinforce each other and drag the system further down. The most serious consequence is found in the failure of modern agriculture and the resulting malnutrition. The collapse of infrastructure not only makes it more difficult to deal with the decline of agriculture and other immediate disasters, but acts to amplify the crisis and leads to further social disintegration.

The various far-flung impacts and the numerous interlinking problems render the crisis nearly impossible to remedy. Even with a healthy economy, it is doubtful that North Korea could now repair its degenerated society. Though the original problem may have been a lack of fuel, it cannot be corrected now by a simple increase in fuel supply. At this time, it will take an international effort to save the people of North Korea. And given the current political animosity between DPRK and the US, it is doubtful that this international effort will take place.

The painful experiences of DPRK point out that dealing with an energy crisis is not just a matter of finding an alternative mode of transportation, an alternative energy source, or a return to organic agriculture. We are talking about the collapse of a complex system, in this case a social system which evolved gradually from a labor intensive agrarian society to a fossil fuel-supported industrial, technological society. It simply is not possible to step back to an agrarian society all at once, or to take a leap forward into some unknown high-tech society. Complex systems change gradually, bit by bit. Faced with immediate change, a complex system tends to collapse.

For a world facing the end of growing energy production, this means that the changes should have begun decades ago, giving time for a gradual transition. We had our warning back in the 1970s, when there might have been time to make a transition to a society independent of fossil fuels. Now it is simply too late. It is a waste of our time talking about a hydrogen future, or zero point energy, or a breakthrough in fusion. Even if we could find a quick technological fix, there is no time left to make the transition. Our only hope now lies in a grassroots effort to change to a sustainable system, and that is where we should direct our efforts. As anyone familiar with wilderness survival knows, when you give up hope, you give up.

The Next "Green Revolution":
Cuba's Agricultural Miracle

THERE MAY NOT BE time left to change our entire technological civilization, but there might still time to shift into a more sustainable food system. And along the way, we might buy ourselves enough time to convert to a localized, equitable, and sustainable economy. A sustainable food system would entail, but not be limited to, localized agriculture, no-till agriculture, composting, natural and organic farm management practices, farmers markets, urban and community gardens, co-operatives and community supported agriculture (CSAs). While it would be a great help if the government were involved in promoting this transition by providing incentives for small farms, organic farmers, urban gardening, and localized markets, the transition could succeed with grassroots participation alone. Before we discuss some of the solutions to the coming crisis in industrial fossil fuel-based agriculture, let us look at an example of a country that lost its energy base but was able to develop, instead, a more sustainable form of agriculture.

The collapse of the former Soviet Union left Cuba in much the same situation as the DPRK. When the Soviet Union collapsed, Cuba lost its oil imports and its major trading partner. US sanctions kept the country isolated.

However, there are some very important differences between Cuba and DPRK. For one thing, Cuba has a much warmer climate, with a longer growing season. Cuba also has a better ratio of population to arable land, though most of the arable land is not of the best quality.[1] Cuba has a large percentage of scientists, engineers and doctors in its population. With only 2 percent of Latin America's population, Cuba holds 11 percent of the region's scientists.[2] Even before the crisis provoked by the collapse of the Soviet Union, Cuban scientists had begun exploring alternatives to fossil fuel-based agriculture. Research into ecological agriculture began back in the 1980s. By the time of the crisis, a system of regional research institutes, training centers, and extension services was in place to quickly disseminate information to farmers.[3] And finally, the Cuban government had social programs in place to support farmers and the population through the crisis and the transition into ecological agriculture.

Before looking at the crisis and the Cuban response, it is necessary to look briefly at Cuban society before the crisis, particularly rural society and the agrarian reforms of past decades. It is here that the groundwork was laid for a successful transition.

A Short History

Before the 1959 revolution, there was one word to describe Cuba: inequality. Only 8 percent of the farmers controlled 70 percent of the land. US interests controlled most of the economy, including most of the large plantations, a controlling interest in the sugar production, the mining industry, oil refineries, electrical utilities, the communications system, and many of the banks.[4]

The majority of the rural labor force consisted of landless, seasonal workers without schooling, healthcare, electricity, or running water. They earned their living during only three months of the year — at planting time and at harvest. Rural workers were lucky to earn one-quarter of the national income.[5]

At the time of the revolution, most of the wealthy landowners fled to the United States. Their former holdings were expropriated and given over to the laborers. The Cuban revolution has been followed by three periods of agrarian reform, first in 1959, secondly in 1963, and finally the current land reform of the 1990s.

The first reform limited private land owning to 1,000 acres. This tripled the number of small farmers and replaced the large plantations with state farms. The second agrarian reform further limited private land ownership to 165 acres per person.[6] The land reform of the 1990s is more properly called a controlled privatization. We will discuss it later.

By 1965, state farms controlled 63 percent of the arable land, and over 160,000 small farmers owned and worked an additional 20 percent.[7] The small farmers joined farmer associations, Credit and Service Cooperatives (CCSs) and Agricultural Production Cooperatives (CPAs), which together controlled 22 percent of the arable land. The CCSs and CPAs are in turn confederated in the National Association of Small Producers (ANAP), which provides training and a number of services to its members and negotiates with the government for prices and credit. ANAP members produce 52 percent of the vegetables grown in Cuba, 67 percent of the corn, and 85 percent of the tobacco.[8] Another 20,000 small farmers own their land independently of cooperatives. These unaffiliated private farmers own about one percent of the arable land.[9]

The agrarian reforms succeeded because the government was truly intent on redistributing wealth and creating a more equitable society. Farmers and cooperatives were supported with low-interest credit, stabilized prices, a guaranteed market, technological assistance, transport, and insurance. The government also enacted laws which prevented the reconcentration of land, effectively blocking former plantation owners from slowly buying back their estates. The revolution took back control of Cuba from the US through laws banning foreign ownership of property. Cuba's isolation, in fact, had some positive benefits: it allowed Cubans to realize their social transformation without outside intervention. And finally, the population was educated and provided with decent health care.

By the 1980s, Cuba had surpassed most of Latin America in nutrition, life expectancy, education, and GNP per capita. The literacy rate was an astonishing 96 percent, and 95 percent of the population had access to safe water.[10] Cubans achieved a large degree of equity and industrialization through a trade regime that was highly import-dependent. Importing transportation vehicles, machinery, and fuel allowed them to build up their industrial base, which they did in a more equitable manner.

From the time of the revolution to the 1980s, Cuban agriculture became more mechanized than any other Latin American country's. By the end of the 1980s, state-owned sugar plantations covered three times more farmland than did food crops. Sugar and its derivatives made up 75 percent of Cuba's exports, sold almost exclusively to the Soviet Union, Eastern and Central Europe, and China.[11]

However, because Cuban agriculture was overwhelmingly dedicated to sugar, tobacco, and citrus, the country had to import 60 percent of its food, all from the Soviet bloc. Cuba also imported most of its oil, almost half of its fertilizer, 82 percent of its pesticides, 36 percent of its animal feed for livestock, and most of the fuel used to produce sugar.[12] This system of imports and exports allowed Cuba to modernize and raise the standard of living and quality of life for all its residents. However, its dependence on the Soviet Union and the focus on sugar production left the country extremely vulnerable should anything happen to its major trading partner.

Crisis

The first few years after the Soviet Union collapsed had a severe impact on Cuba. Almost overnight, Cuba lost 85 percent of its trade. Fertilizer, pesticide, and animal feed imports were reduced by 80 percent.[13] Imports of fertilizer dropped from 1.3 million tons per year to 160,000 tons in 2001. Herbicide and pesticide imports dropped from a combined 27,000 tons to 1,900 tons in 2001.[14] And petroleum supplies for agriculture were halved.[15]

Food imports, which had accounted for 60 percent of Cuba's food, were also halved.[16] And by 1994, agricultural production had dropped to 55 percent of the 1990 level.[17] Per capita daily caloric intake dropped from 2,908 calories in 1989 to 1,863 calories in 1995, a decrease of 36 percent. Protein intake decreased by 40 percent,[18] and dietary fats dropped 65 percent.[19] There are estimates that from the collapse of the Soviet Empire to 1994 the average Cuban lost 20 pounds.[20] Undernourishment jumped from less than 5 percent to over 20 percent, the largest increase in all of Latin America during the 1990s.[21]

The crisis was compounded by the US, which tightened its already stringent economic blockade. Throughout the worst years of the crisis, 7,500 excess deaths per year can be directly attributed to the US sanctions.[22]

Two government policies are credited with preventing the crisis from becoming catastrophic: food programs targeting particularly vulnerable populations (the elderly, children, and pregnant and lactating mothers), and the food distribution ration card which guaranteed a minimum food provision for every citizen (albeit greatly reduced from former levels). This government-maintained safety net kept the crisis from reaching depths comparable to North Korea, while giving the country breathing space to redesign its agricultural sector to meet the challenge.

The agrarian reforms of the mid-1990s were the key to recovering from the food crisis, but they could not have worked without the earlier agrarian reforms and without an educated and modernized peasantry unique in Latin America. The Cuban miracle is the product of a people with vision and solidarity.

The Cuban Miracle

Cuba's economy is now recovering from the loss of its closest trading partner. Cuban GNP has grown every year since 1995. There have been solid gains in employment, productivity, and exports. Fruit production has returned to its 1989 level (and even surpassed it in the case of plantains). Vegetables and tubers for domestic consumption have seen a prodigious increase in production. Daily food intake has climbed to 2,473 calories per person, a 33 percent increase over caloric intake in 1994.[23] Observers the world over have pronounced the Cuban efforts a success. Single-handedly, without help from either the World Bank or the International Monetary Fund (and in total contrast to the normal

World Bank and IMF reform policy), Cuba has disproved the myth that organic agriculture cannot support a modern nation. Agrarian reform in the 1990s centered on creating a new system of sustainable agriculture, developing healthy markets, and converting the unwieldy state farms into worker-owned cooperatives.

For decades, Cuban scientists had been aware of the negative effects of industrialized agriculture. Soil erosion and mineral depletion had been a marked problem. Before the crisis of the 1990s, scientists had already developed organic and ecological methods of farming. Following the crisis, the Cuban government embraced these new methods and promoted them with new agrarian policies. By comparison, in North Korea there was little interest in organic agriculture before their crisis. North Korean science and technology were dedicated to industrialization, just as the government was dedicated to militarization.

The task was to convert the nation's agriculture from high-input, fossil fuel-dependent farming, to low-input, self-reliant farming. Farmers did this by first remembering the techniques that their ancestors had used before the advent of industrial agriculture, like intercropping and manuring. Secondly, farmers used new environmental technologies offered as the result of scientific development, such as biopesticides and biofertilizers. Biopesticides — using microbes and natural enemies to combat pests — were introduced along with resistant plant varieties, crop rotation, and cover cropping to suppress weeds. Biofertilizers were developed using earthworms, compost, natural rock phosphate, animal manure and green manures, and the integration of grazing animals. To replace tractors, farmers returned to animal traction.

BASIC UNITS OF COOPERATIVE PRODUCTION (UBPCs)

The large state farms were incompatible with this new paradigm. Agroecological farming simply does not work on a large farm. In industrial farming, a single technician can manage thousands of acres without intimate knowledge of the land he is overseeing. A few random observations will provide him with all the input he needs to write out instructions for the application of a particular fertilizer formula or pesticide to be applied with machinery over the entire area. However, in agroecological farming, the farmer must be intimately familiar with every patch of soil, knowing exactly where to add fertilizer, and where pests are being harbored or are entering the field. Smaller farms were easier to manage, and more compatible with sustainable agriculture.

In September 1993, the government instituted a new program to restructure state farms as private cooperatives owned and managed by the workers. These new cooperatives were called Basic Units of Cooperative Production (UBPCs). The new program transferred 41.2 percent of the arable land — most of the state farms in the country — into 2,007 new cooperatives with a membership totaling 122,000 people.[24] To link the workers to the land, the cooperative owns the crops and members are compensated based on their productivity, not their timesheet. This provides a greater incentive within the cooperative, yet allows it to offer larger economies of scale, mechanization, and collectivist spirit.[25]

AGROECOLOGY

Agroecology is the science of applying ecological concepts and principles to the design, development, and management of sustainable agricultural systems. It is a whole systems approach to agriculture that embraces environmental health and social equality, while balancing these with economic viability. The goal is long term sustainability for all living organisms, not just humans.

Agroecology is guided by many principles, but these can all be placed under the following broad categories:
- Use renewable resources
- Minimize toxic inputs and polluting outputs
- Conserve resources
- Manage and re-establish ecological relationships
- Adjust to local environments
- Diversify
- Empower people
- Manage whole systems
- Maximize long-term benefits
- Value the health of all living things and ecosystems

For more information, visit http://www.agroecology.org/

Although the government retains ownership of the land, the UBPCs are granted a free lease to it. The government then contracts with the UBPCs for the crops and amounts to be grown. On the basis of these contracts, the government sells the necessary agricultural inputs to the UBPCs.

The new system has not been without problems. Most notably, there is friction between the UBPCs and local officials of the Ministry of Agriculture who still behave as though they are in control of the cooperatives. However, the trend is clearly heading toward greater autonomy for the cooperatives.

PRIVATE FARMING

The holdings of private farmers have also grown in the last decade. Since 1989, the government has turned over nearly 170,000 hectares of land to private farmers.[26] Although the government retains title to the land, private farmers and CPAs can farm it rent-free for an indefinite period of time. As a result, many Cubans now view farming as an opportunity and many families have left the cities to become farmers. The ANAP claims that its membership expanded by 35,000 from 1997 to 2000.[27] The new farmers tend to be young families (many of them college educated), early retirees, or workers with a farming background.

The Credit and Service Cooperatives (CCSs), made up of small, independent farmers, have outperformed the CPAs, the UBPC cooperatives and the state farms. And this achievement has come despite limited credit. As a result, the ANAP began a program in 1998 to strengthen the business side of the CCSs. CCS cooperatives are now allowed to open bank accounts, hire market representatives, and plan collectively. Once qualified as "strengthened," a CCS gains the ownership of machinery and the ability to collectively market the goods of its members.[28]

URBAN AGRICULTURE

Another bright spot in the reforms is urban agriculture, though this originated as a spontaneous development which was later backed by policies. Today, half the produce consumed in Havana is grown in that city's urban gardens. Overall, urban gardens produce 60 percent of the vegetables Cubans consume. Urban farmers grow the equivalent of 215 grams of vegetables per day per person for the entire population. [29]

Neighborhood gardens and community horticultural groups not only produce food for their members, they donate produce to schools, clinics, and senior centers, and still have enough excess produce to sell in the neighborhood. Neighborhood markets sell produce at well below the cost of the larger community markets, providing fresh vegetables for those who cannot afford the higher prices. By the beginning of the year 2000, there were 505 vegetable stands in Cuban cities, with prices 50 to 70 percent lower than at farmers markets.

Recognizing the potential of urban agriculture, in 1994 the government created an urban department in the Ministry of Agriculture. The Urban Agriculture department formalized the growers' claims upon vacant lots and legalized their rights to sell their produce. The department has acted to support and promote urban agriculture without attempting to impose its authority upon the movement. Laws require that urban produce be completely organic, and ban the raising of livestock in urban areas. All residents can claim up to one-third of an acre of vacant land on the edge of the major cities. By the beginning of the year 2000, more than 190,000 people had applied for and received these personal lots.[30] The government has also opened a number of neighborhood agricultural stores to supply organic inputs and extension services.

Gardeners are empowered by working to provide food for themselves and their neighbors. As one urban gardener said, "We don't have to wait for a paternalistic state to do things for us. We can do it for ourselves."[31]

There are many diverse forms of gardening referred to collectively as urban gardening. The most common are *organóponicos*, which farm raised beds of organic material, utilizing biological pest control and organic fertilizer. Some *organóponicos* even have micro-jet irrigation and mesh shading. *Organóponicos* are highly productive, yielding anywhere from 6 to 30 kilograms of produce per square meter.[32]

AGRICULTURAL MARKETS

In October 1994, the Cuban government opened 121 agricultural markets throughout the country.[33] As an immediate consequence, the black market in basic food items virtually disappeared. Food prices in the open market were a good deal less than on the black market. The free markets also quickly demonstrated that they led to increased production and spurred higher quality and greater diversity in produce.

Over time, however, supply and demand pricing did result in rising food prices. By the year 2000, food purchases could take up as much as 60 percent of the average Cuban salary. The poor and the elderly turned to urban vegetable stands offering produce from urban gardens.

Studies have shown that the major culprit in rising market prices were the distributors. The lack of fuel in Cuba has resulted in severe transportation shortages. The few people who did own trucks colluded to pay little to the farmers and then charge high prices to the vendors. Some distributors have gained profits of as much as 75 percent.[34]

To combat this problem, the Ministry of Agriculture is giving used trucks to private cooperatives to allow them to bypass the distributors and take their goods directly to market. The remaining state farms are also selling their produce at low prices in state agricultural markets, in an effort to drive down prices. The experiment in free agricultural markets has shown Cubans that there must be some government controls on price gouging and collusion.

Results

Though caloric intake has not yet reached the levels of the 1980s, few would dispute that domestic food production in Cuba has made a remarkable recovery. During the 1996–1997 growing season, Cuba attained its highest ever production level for ten of the thirteen basic items in the Cuban diet.[35] And in 1999, agriculture production increased by 21 percent over the previous year.[36] Comparing food production to 1989 levels is not quite so favorable, but still impressive.

Animal protein production still remains close to depressed 1994 levels. This is partially because the market reforms cannot apply to meat, eggs and milk, which are not easily sold in farmers markets, due mostly to lack of refrigeration. Likewise, the agroecological model is not so easily applied to animal production. But the biggest factor keeping animal protein production down is that the transition from industrial animal breeding to sustainable, ecologically feasible animal breeding must proceed at a much slower pace than the similar transition in agriculture. Factors slowing the transition in animal breeding include waste disposal, disease control, and humane treatment of the animals.

Exports are still considerably lower than 1989 levels. Only citrus exports have regained that level. Coffee and tobacco exports still lag behind, and sugar exports are only a fraction of 1989 levels.[37] In the case of sugar production, the US embargo and the low price of sugar on the world market are keeping production depressed. But the Cuban government is formulating plans to increase sugar exports in an attempt to bring in much-needed foreign revenue and investment.

Aside from restoring export levels and animal protein production, the future of the new Cuban agricultural model faces three major challenges: reconciling price distortions between the US dollar and the Cuban peso, reconciling state control and private initiatives, and overcoming limits to the ecological model. Concerning this latter challenge, agroecological farming requires more land and more labor than industrial farming. While Cuba does have the land base to continue agricultural expansion, rural areas have experienced a labor shortage. Only 15 percent of the Cuban population lives in the countryside.[38] The agricultural sector has been able to reverse the rural-to-urban migration and attract the necessary workforce, but nobody is certain how long this reversal will continue. And then there is the uncertain balance between farm labor requirements, the higher caloric intake necessary for busy farmhands, and agricultural production.

The new Cuban model of agriculture faces many challenges, both internally and externally, but that does not diminish its current success. And there are many analysts who feel that the Cuban experiment may hold many of the keys to the future survival of civilization.

Cuba's Example

The World Bank has reported that Cuba is leading nearly every other developing nation in human development performance. Senior Bank officials have even suggested that other developing countries should take a closer look at Cuba.[39] This is despite the fact that the Cuban model flies in the face of the neoliberal reforms prescribed by both the World Bank and the IMF. Because Cuba's agricultural model goes so against the grain of orthodox economic thought, the World Bank has called Cuba the "anti-model."

Indeed, Cuba's fastest growing export is currently ideas. Cuba regularly hosts a number of visiting farmers and agricultural technicians from throughout the Americas (excluding the US), and elsewhere. Cuban agriculture experts are currently teaching agroecological farming methods to Haitian farmers. Ecologists as well as agricultural specialists are finding great promise in the idea that biodiversity is not just a conservation strategy, but a production strategy too.

As declining fossil fuel production impacts civilization, Cuba may find itself in a position to help lead the world into sustainable agriculture. Currently, few countries are willing to invest in human capital and infrastructure the way that Cuba has, but hopefully this will change in the years ahead.

Resistance to Cuban-style agricultural reform would be particularly stiff in the United States. Agrobusiness will not allow all of its holdings and power to be expropriated. Nor is the US government interested in small farms and organic agriculture. The direction of US agriculture is currently towards more advanced technology, greater fossil fuel dependency, and less sustainability. The ability of small farmers and urban gardens to turn a profit is effectively drowned out by the overproduction of agribusiness.

However, now is the time for people to study agroecology (and permaculture as well), with an eye towards implementing this technology once declining fossil fuel production sparks a crisis in industrial agriculture. Our survival will depend upon our ability to implement these ideas once the current technology has failed. The North Korean example shows that the alternative is unthinkable, and the Cuban example shows what is possible.

8

Building a Sustainable Agriculture

Food Security

Cuba and North Korea are both special cases. These countries lost their energy base in the course of a few short years. The result was an instant crisis. The decline of world energy production will happen at a much slower pace, and that pace could offer the opportunity to make a successful transition to a more sustainable food system.

It is a question of the rate of decline and how the world reacts to that decline. A rapid decline of seven percent or faster will not allow us an opportunity for transformation. However, a decline of two percent per year (which is what many experts expect) could be accommodated by a transition to a stable-state economy, sustainable agriculture, and a natural decline in the world's population with a minimum of suffering. This is barring an economic panic or global resource wars.

There is already a global movement towards sustainable agriculture. In order to make the transition, this movement needs wide support. We all need to introduce sustainable practices into our lives and into our communities. What is required is a complete transformation in the way we interact with our environment, not just on our farms, but in our cities and suburbs, and throughout the whole range of human-environment interaction. We need to integrate food production back into our daily lives, by gardening, foraging, and animal husbandry. Farms need to practice sustainable agriculture and should be recognized for the vital service they could and should be providing to surrounding communities.

Food security is a matter of homeland security. To have food security, especially in the face of emergencies, a community should be able to provide 30 percent of the food required by its residents. The rest would come from neighbouring farms. Presently in the US, less than five percent of food is produced locally.[1] To provide food security, we need a localized system of food distribution from farms practicing sustainable agriculture, in combination with urban farming and permaculture.

Sustainable Agriculture

The practices that are referred to as sustainable agriculture are numerous and vary depending on the local ecology. However, a number of general practices are more or less common to all sustainable agriculture systems. These are crop rotation, cover crops, no-till or low-till farming, soil management, water management, crop diversity, nutrient management, integrated pest management, and rotational grazing. While sustainable farming does not always engender organic farming, for the purposes of this discussion we will assume that it does. Organic farming is the most practical method of reducing fossil fuel input at the level of production.

A number of articles and organizations make bold claims about organic farming, stating that it could feed the entire current human population and provide increased production over present methods of industrialized agriculture. These claims appear to misinterpret the data.

Studies have shown that sustainable agriculture can produce an average 93 percent increase in food production per hectare.[2] However, this increase diminishes as the size of the farm increases. At some unknown critical scale, industrial farming becomes more efficient. As a rule, organic crops only require 50 percent of the energy input per unit area that conventional crops require. This savings comes from the reduction of artificial fertilizers and pesticides.³ But when the energy input is calculated per unit of output rather than unit of area, the lower yields of large-scale organic farming reduce the energy ratio. Here too, the larger the operation, the larger the energy costs with regards to planting and harvesting.

Sustainable agriculture, for all intents and purposes, means a return to small-scale farming, where the acreage can be managed by a family, and a horse or mule with a plow. John G. Howe, a retired engineer, has developed a ten horsepower solar-powered tractor that uses no gasoline and could provide much of the mechanical energy needed on a small farm.⁴ His book, *The End of Fossil Energy and the Last Chance for Sustainability,* offers a lot of excellent advice. However, the dependence of solar technology on fossil fuels for manufacturing energy and feedstocks, along with the ramping up of solar cell production necessary for mass production are problematic.

In comparing farm systems with natural ecosystems, we quickly see that the basic problem with conventional agriculture is that it is not a closed system. In nature, nutrients (most importantly nitrogen, phosphorus, and potassium) are cycled through the system. Nutrients in an ecosystem flow from the physical environment through plants to animals. Then plant and animal waste, along with dead organisms, are decomposed by fungi and microbes, and returned to the physical system where they begin the cycle all over again. True, some nutrients are lost to the local ecosystem through runoff and by other means, but the amount of these losses is small and is usually replaced by the process of breaking down bedrock and soil formation.

In our conventional agricultural system, we bring in massive amounts of nutrients from outside of the ecosystem. In our efforts to maximize production, we abuse the ecosystem with a flood of artificial and imported inputs. Then we ship the products out of the ecosystem to be consumed by humans and ultimately to be disposed of in landfills and sewage systems. Conventional agriculture is a gigantic through system that depletes our resources, exhausts our farmlands, and results in overwhelming mountains of garbage.

For a sustainable agriculture, and a sustainable society, we need to close the loop and integrate agriculture with our settlements, reestablishing the cycle of nutrients. Swedish engineer Fölke Gunther points out that two changes are needed for agriculture to mimic ecosystem nutrient circulation:

- Animal feed should be produced on the same farm as the animals, and animal wastes should be returned to the land where the animals are fed, either directly or through composting and effective mixing into the soil.
- Through the use of source-separating toilets, nutrients exported as human food should be collected and returned to the farmland uncontaminated.[5]

More information on the latter practice can be found in *The Humanure Handbook,* by Joseph Jenkins.

Using phosphorus as the critical limiting nutrient required for agriculture,[6] Dr. Gunther has shown that the phosphorus export from 0.4 hectares of balanced agriculture is equal to the phosphorus content of excrement from five to seven people.[7] A balanced agriculture will require between 0.23 and 0.15 hectares to provide the necessary nutrients for one human being. Thus, a 40 hectare balanced farm could support around 200 people.[8]

Scaling up from this point, a community of 800 to 1,200 people living in 3 or 4 connected settlements would require a balanced agricultural area of 160 to 240 hectares. Adding in areas for the improvement of local ecosystems and recreation would increase the land required to 170 to 260 hectares.[9] A settlement of this size could provide for the cultural and social needs of the population, provide social diversity, and allow for the variety of necessary skills and employment.

Large scale implementation of balanced and integrated agriculture would require a planned ruralization, where a portion of the urban population would move into balanced settlements in the hinterland of the urban areas. Cuba experienced a similar rural migration of former city dwellers as a part of its agricultural experiment. In a society converting to sustainability, this migration from the cities would be as natural a phenomenon as the urbanization of the past century was a natural phenomenon of industrialization. The remaining urban population would need to develop an integrated agriculture within their own cities, along with a system for recycling their wastes.

Urban Agriculture

Urban food production is already a growing phenomenon throughout the world. In 1996, urban farms produced 80 percent of the poultry and 25 percent of the vegetables consumed in Singapore.[10] As a result of the failure of the Soviet Union, between 1970 and 1990, the number of Moscow families engaged in food production rose from 20 to 65 percent.[11] Currently, 44 percent of Vancouver, British Columbia's population already grow some vegetables in their gardens. There are over 80,000 garden allotments along railroad tracks and elsewhere in Berlin, and the list of those waiting for further allotments to become available is nearly endless.[12] In the US, counties defined as urban or urban-influenced grow 52 percent of the dairy products, 68 percent of the vegetables, and 79 percent of the fruit consumed there.[13]

The potential for urban gardening is enormous. Community gardens would be an excellent use of abandoned inner city areas. Limited leases to abandoned lots could allow gardeners to produce immediate benefits from land that ordinarily lies vacant for an average of 20 to 30 years. Instead of being magnets for litter, rats, and crime, such lots could become showplaces and centers for community socialization.

Other unused city lands that could be converted to agriculture include portions of parks, utility right of ways, roadway medians and center dividers, and unused school and hospital grounds. Hospitals could keep their food bills down while adding healthy fresh produce to their patients' diets. Likewise, schools could grow produce for their lunch programs, while giving their students a firsthand opportunity to learn about plant growth and farming.

There is one major unused surface area in cities that is well suited to container gardening: rooftops. On the average, rooftops comprise 30 percent of a city's total land area,[14] and rooftops enjoy the full benefit of sunshine and rainfall. Rooftop gardening could provide a substantial portion of urban dwellers' food.

Taking this theme a step farther, some city streets could be closed and converted to gardens and orchards. Likewise, some of the streams that have been channeled into sewers could be brought back to the surface, remediated, and even used for irrigation. If agriculture were fully integrated into urban life, it could conceivably produce a revitalized and verdant environment, where communities are interlaced with gardens, orchards, parklands, and open waterways.

Community gardens have proven that they can not only feed their members; they can also provide a source of fresh produce for food kitchens, food banks and other food assistance programs. Meanwhile, home gardens make use of private yards,

COMMUNITY SUPPORTED AGRICULTURE

Typically, members or "share-holders" of the farm or garden pledge in advance to cover the anticipated costs of the farm operation and farmer's salary. In return, they receive shares in the farm's bounty throughout the growing season, as well as satisfaction gained from reconnecting to the land and participating directly in food production. Members also share in the risks of farming, including poor harvests due to unfavorable weather or pests. By direct sales to community members, who have provided the farmer with working capital in advance, growers receive better prices for their crops, gain some financial security, and are relieved of much of the burden of marketing.

From "Community Supported Agriculture (CSA): An Annotated Bibliography and Resource Guide," Suzanne DeMuth.

www.nal.usda.gov/afsic/csa/csadef/htm

Community Supported Agriculture (CSA) programs directly link local residents and nearby farmers, eliminating "the middleman" and increasing the benefits to both the farmer and the consumer. In a CSA program, a farmer grows food for a group of local residents (called "shareholders" or "subscribers") who commit at the beginning of each year to purchase part of that farm's crop. The shareholders thus directly support a local farm and receive a low-cost weekly or monthly supply of fresh, high-quality produce. The farmers receive an initial cash investment to finance their operation and a higher percentage of each crop dollar because of direct delivery. Both parties jointly share the benefits and risks.

From "Community Supported Agriculture," James Wilkinson.

www.rurdev.usda.gov/ocd/tn/tn20.pdf

decks, balconies, rooftops, and even windowsills and indoor aquariums to produce everything from tomatoes to honey to small livestock and fish.

Farmers Markets and CSAs

Localized agriculture for all practical purposes requires the revitalization of farmers markets. Farmers markets allow farmers to reconnect to local communities, and allow residents to reconnect to the source of their food. They also cut the middleman out of the food system, which is where the profits of commercial agriculture have been increasingly concentrated over the past century.

Farmers markets also reduce the energy costs of food distribution. A study in Toronto, Canada, found imported foods purchased through the conventional food system traveled 81 times farther than locally produced foods found at the farmers market.[15] Another study showed that if Iowa produced just 10 percent of the food it consumed, it would save 280,000 to 346,000 gallons of fuel.[16]

CSAs represent another option for localized agriculture. CSA is short for Community Supported Agriculture. In a CSA, a community or group of individuals pledge their support for a farm operation and the farm, in return, shares its produce with the CSA members. Members undertake a legal and/or spiritual obligation for the maintenance and continuing prosperity of the farm. CSA members are given an opportunity to connect personally with the source of their food, and may even participate directly in its production. In exchange, the farmer receives capital backing, better prices for his or her crops, financial security, and freedom from having to market her or his produce.

9

Twelve Fun Activities for Activists

THE MOST LOGICAL and ethical thing for governments and corporations to do in the face of a new era of energy depletion would be to limit their energy consumption so as to begin conserving before the fact. Unfortunately, logic and ethics have little to do with how the world is run.

It is naïve to think that government and corporate decision makers are genuinely concerned with the common good and would be willing to temper current economic production to offset some apparently distant problem. Numerous historical examples from just the last couple of hundred years, since the development of modern representative democracy, illustrate that the system does not work like this.

The job of government is to promote business as usual. Politicians don't want to focus on anything beyond the next election, particularly if the solution might hinder the economy. Likewise, corporations are only interested in their immediate profits. This is a very shortsighted system all the way around. It is more profitable for all of them, at this moment, to maintain their state of denial.

Our supposed decision makers really prefer to lead from the rear. Every social and environmental advance that has ever occurred started first by building a grassroots movement. Awareness must first take root in the general population, building a popular consensus and demand for action before our supposed leaders will climb on the bandwagon to take the credit. And even then, the public must be wary that our leaders do not enact partial and ineffective measures simply to placate the population.

Before we can ever hope for federal solutions, those of us who are aware of these problems need to reach the public and educate them about the reality of energy depletion. We need access to the media, and we need the support of powerful environmental organizations or — if those organizations are too compromised — we need to develop new organizations for community-based action.

Unfortunately, all of this presents us with an uphill battle. Even though peak oil is now being talked about more than it was a few years ago, the media and the environmental organizations still seem to be lined up against us. If they do mention the energy crisis, then they are quick to tout their favorite solutions, whether it is renewable energy, hydrogen fuel cells, or whatever. And let us not even mention the government and corporate opposition.

There are some who think our governments are conspiring to keep us in the dark while the elite vie for the best position in the upcoming energy crisis. I'm doubtful of a system-wide conspiracy, but I suspect some knowledgeable players are milking the most profits out of the current situation in preparation for what is to come. Witness the profit statements posted by the oil majors. I see nothing to support current energy prices in the latest data. The system is beginning to strain, but oil and natural gas stocks held commercially and by the federal government are well within the five-year average.

Instead of a wide-ranging conspiracy to fleece the general public and then bring the system down around their heads, I think this denial is endemic to our current economic system. It gives government and corporate decision makers little impetus to grapple with the problem of energy depletion, and good reason to cling to business as usual.

Most of our leaders are in fact deluded by their near-religious faith in capitalism and the free market system. They place their trust in technology and the power of the human intellect to overcome all problems. Their devotion and rapturous transcendence are exceeded only by the fundamentalist Christians, with whom many of them are closely allied.

Expecting our decision makers to lead us through this crisis would be like asking the blind to lead the blind. Forget about them; they surely have forgotten about us. Begin talking to your neighbors and building awareness in your community. Organize to create community gardens, community markets, and community bicycle marts. Work with your neighbors on taking your neighborhood off the grid. Develop a local currency or a system of barter.

We cannot afford to play follow the leader. The only thing that is going to see us through this crisis is grassroots community action. And, indeed, this is the only thing that ever has worked for us.

For those who are becoming aware of the great impact energy depletion will have on our lives, the realization that we cannot rely on our so-called leaders to solve these problems can lead to panic and despair. The vast majority of the public hasn't got a clue. If you try to inform them, most don't want to know, while the rest place their faith in our leaders or in a technological breakthrough. How can we ever build a grassroots movement when most people can't even perceive the problem?

Such is the case with all social movements: they start with a small but active group of people who are the first to be aware of the problem and the necessity for change. This is how the Vietnam War protest began, as well as the civil rights movement, the environmental movement, and every other major grassroots movement. They were all started by a handful of people scattered through the general population. Their first steps were to network with those who were aware, organize, and begin to broadcast their message to any who might be listening. One important factor is to provide a visible position that will attract attention as time goes on and awareness spreads.

But energy depletion is different from all those other social problems. In none of them was there a critical point beyond which a solution was no longer possible. This might be the case, however, with energy depletion. The worry is that there *is* a point of no return, where it has wreaked too much havoc with our economic system, our agricultural and food distribution system, and our manufacturing base — a point of no return beyond which the total collapse of civilization can no longer be prevented or even mitigated. This is what lies at the crux of the panic and despair that strikes so many who become aware of the potential consequences of energy depletion. This is certainly the issue that worries me. Have we slipped too far down the road to Olduvai Gorge? Is Richard Duncan correct? Are we to witness the end of technological civilization within our lifetimes?[1]

But if you do believe we have passed the point of no return, or

THE OLDUVAI THEORY

The Olduvai Theory was espoused by engineer Richard Duncan, and was named after the Olduvai Gorge in Africa, where the oldest human remains were found. At its simplest level, the theory states that the planet Earth holds enough energy resources for only one technological civilization to evolve. The life expectancy of this technological civilization will be around 100 years.

The theory is defined by per capita world energy production — the amount of energy produced worldwide divided by the total world population. The data marks out a roughly parabolic curve, encompassing the time period from 1930 to 2030. The upward side of the curve was marked by the spread of technology and electrical infrastructure. The downside of the curve will be marked by rising energy prices and increasingly severe shortages and blackouts. The theory proposes that eventually humanity will return to a style of life that is local, tribal, and solar, without the frills and comforts of modern technology.

Richard Duncan charts world per capita energy production as rising at 3.45 percent per year from 1945 through 1973. From 1973 through 1979, production slowed to 0.64 percent. Per capita production peaked in 1979. From 1979 until this writing, per capita energy production has remained fairly constant. This plateau is expected to continue until around 2008, after which time it will be followed by a steep decline. The decline in per capita energy production will be closely mirrored by a decline in population. By the year 2030, per capita energy production will have fallen back to where it was in 1930, and Industrial Civilization will have ended.

For a more detailed account of the theory, see "The Olduvai Theory; Energy Population and Industrial Civilization," Richard Duncan, *The Social Contract*, Winter 2005–2006.
www.thesocialcontract.com

soon will pass it, then what are you doing here right now? You had best head for the hills and hope that the folks already in those hills will still welcome you. If you hold no hope for a transition to a sustainable society, then you had better learn to survive on your own, build your little hideaway, and prepare to fight off the starving masses once they sniff you out.

The fact that you are reading these words suggests that you still hope there is a chance for converting to a sustainable civilization. And so long as we can entertain such a hope, there is still a chance. There is a lot of talk of genetic determinism, and the collapse of complex systems. These are interesting philosophical exercises, but in the end they are just intellectual excuses for giving up on humanity. Any person with survival experience can tell you this: you give up your chance to survive when you give up hope and stop trying. When you give up hope, you close the door on a sustainable transition.

But it is questionable whether hope alone is enough. You have to become active; you have to put forth effort. Do not be dissuaded by negligible results; keep on trying. The secret to movements is that they grow exponentially. For a long time, it will seem that you are making hardly any progress at all, and then you suddenly find yourself swamped by a flood of community awareness.

This is what we have to plan for at present. We have to organize ourselves and prepare for the day when we are nearly overrun by all of the people who suddenly see the problem. And we must hold faith that this day will not arrive too late.

For now we need to organize. Start hosting energy depletion or sustainability awareness gatherings in your house. Talk to your family, your friends and your neighbors. Attend rallies and provide a visible presence. Carry signs, pass out fliers. Attend sustainability and alternative energy conventions; be critical but call attention to the need for low-tech, grassroots solutions — even though they are clearly only partial solutions. Become a familiar face at your local farmers market; join or start a community garden. Be involved. Prepare yourself for a transition to self-sufficiency, and at the same time be ready for the flood of attention that will come when energy depletion can no longer be ignored. Here follows a short list of activities that activists might engage in to strengthen local food security and help to prepare their community for a sustainable future.

Be involved and never give up hope.

Community Vegetable Gardens

Lobby your communities and neighbors to allow you to plant and tend vacant lots. If you live in an apartment complex with a suitable roof, lobby the management to allow you to build and tend planting boxes on the roof. This is an activity which can foster a strong sense of community between you and your neighbors. You could sell your excess produce at a discount to the local food bank or soup kitchen. Or you could use the excess produce to start a food bank or soup kitchen.

Operation Johnny Appleseed

Take a cue from that activist of American folklore and just start planting. Always save your apple seeds, pear seeds, peach pits, grape pips, cherry pips, etc. Save them and plant them wherever you find a likely spot. You can do this with any hardy perennial — fruits, roots, and a select group of vegetables. You can even donate a little money and time to the purchase and planting of saplings, vines (grapes), bushes and brambles (blueberries and raspberries), or runners (strawberries).

You could also take a tip from the American Indians and other native peoples and promote the propagation of beneficial wild plants. The study of permaculture lends itself to this activity.

Food Not Bombs

This is perhaps the best single idea to come out of the Anarchist Movement in the last fifty years. What is wrong with the Salvation Army, soup kitchens, and other charities? For one thing, the majority of food charities expect something in return, usually a religious conversion. But there is a more basic problem with traditional charities: they are charities. People who are well off are taking time to help the downtrodden. However well-meaning, those who come for the food are made to feel like beggars, beholden for the charity they receive and dependent on the charity of others.

In Food Not Bombs, fliers are passed out announcing an open picnic at a local park, or some such place. The food is prepared ahead of time and laid out where everyone can serve her or himself. And then everyone sits down to eat together.

Food Not Bombs picnics can be combined with educational tours to identify local edible plants. You could even show off the fruits of your free plantings or invite folks to help out with the local community garden.

Farmers Markets and CSAs

If you can locate a nearby farmers market or Community Supported Agriculture (CSA), then patronize the former or join the latter. CSAs are farms dedicated to serving their subscribers. For a yearly subscription price, members are given a percentage of the produce. Subscribers sometimes pitch in with the harvesting and other activities. The farmers usually accompany their produce with advice for food preservation.

If you can't find a local farmers market or CSA, then perhaps you should consider organizing one.

Community Transportation Networks

This is an idea which is bound to become more popular as gas prices go up. Form a community car pool, not just for the commute to and from work, but to shopping centers and elsewhere. You could set up a local network to match up people who need to go to specific places at specific times, so that they can share rides.

With the US becoming increasingly dependent on foreign oil, and with US soldiers dying in oil wars, isn't it unpatriotic for each single person to drive around by him or herself?

Bicycle Co-ops and Bicycle Trails

Bicycle co-ops could maintain a fleet of bicycles for the use of members, or for temporary rental by non-members. The co-ops would maintain the bicycles, and perhaps collect them and return them to distribution centers. Bicycle co-ops could lobby local communities for bicycle trails, and perhaps donate time to the maintenance of those trails.

Support Local Businesses, Particularly Co-ops
How many local businesses are left in the wake of globalization? While supporting local businesses, press to ensure that those businesses are ethical and responsible.

Form Co-ops
There is no end to the essential services which could be provided through co-ops. Co-ops give a community control over the provision of necessities.

Organize Community Activities
Community entertainments such as barn dances, music and art festivals, or community theatres not only provide entertainment, they provide venues where people can socialize. This is where you can meet like-minded folks who would be interested in taking part in the other activities mentioned here. These concerts and festivals also provide forums for local artists to reach an audience and inspire them with visions of where they can take their community.

Other activities provide a pleasant setting for doing tedious work, or group support to get things done. This includes quilting bees, sewing circles, or fix-it fairs where everybody could bring old appliances to fix or salvage.

Community Refurbishing Co-ops
Such groups could help to remodel homes for greater energy efficiency, build or renovate community centers, or possibly build shelters for the homeless.

Community Energy Production Co-ops
Such organizations could provide local, community-owned and maintained, low-level energy production. Depending on local conditions, potential power sources could be wind turbines, solar cells, hydroelectric, or even geothermal.

Ecovillages
Here is the ultimate activity, an entirely self-sustaining community. This is the eventual goal which we must all direct our activities towards if we are to have a free, equitable, and just society. That is, a society where the quality of life makes life worth living and where we can reside happy and contented to watch our children grow up in a positive and healthy environment.

Conclusion

Modern agriculture has charted a course for disaster. Our soils and water resources, and our weakened food crops, will fail us just as energy depletion makes it increasingly difficult to make up for these deficiencies through artificial means. The fossil fuel-based agriculture that allowed our population to climb so far above carrying capacity in the last century will soon falter most cruelly.

There is, however, a chance to transform ourselves into a more sustainable and equitable civilization. So long as the rate of energy production decline is not too high, we could make a successful transition and allow population to shrink back down below the carrying capacity of the planet in a natural fashion, through declines in the birth rate and life expectancy, with a minimum of pain and suffering. But we will all be required to work in order to make this transition happen. We need to redesign our society, aiming for decentralization and localization. We need to reconnect ourselves to the land around us, through home and community gardens, through local small farms and farmers markets, and through permaculture parks and protected wilderness. And we must work hard to organize our communities and awaken our friends and neighbors to this necessity.

It would be much easier if we could depend on our governments and business leaders to make these changes for us, but it is highly unlikely that they will do so in any significant fashion. In fact, the necessary changes will require abandoning the economic and power structures from which these leaders profit. In fact, we should be wary of any government or corporate attempt to "climb on the bandwagon," as it is likely to be an effort to derail and redirect the movement.

As a grassroots movement for relocalization gains support, it is quite possible that it will be met by stiff government resistance. We are, after all, talking about agrarian reform. The US government has a long history of stamping out such movements elsewhere, from the Philippines in the late 1800s, to Guatemala in the 1950s, to Nicaragua and El Salvador in the 1980s,[2] to US attempts to overthrow the democratically elected government of Hugo Chavez in 2002.[3] If a movement of agrarian reform swept through the US, who is to say that the federal government and corporate backers might not become just as violent in stamping it out here.

It is also possible that the elite might use economic disruption to their advantage as a tool to prevent a grassroots transition. It is not beyond the realm of possibility for them to use personal debt to keep the public enthralled. An economic crash, coupled with an inability to escape from personal debt, might prevent us from making the necessary changes. To stop this from happening, perhaps a relocalization movement should also tackle the issue of debt forgiveness.

A SOLARI VENTURE FUND

A Solari Venture Fund acts as both a databank and an investment advisor. As a data tool, the entity charts how resources and financial flows work in a local community. By plotting the money flow within the local community, the data identifies problems (i.e., less than optimal current use of resources in the local community, absence of alignment of incentives, impoverishment within the community). Using this model, members can determine how to restructure their local economy to better benefit their community's overall health.

As a facilitator of equity investment or an investor, a Solari Venture Fund helps plan local re-engineering. It establishes localized investments that attract current capital that is leaving the community, while seeking to attract outside capital as well. A few of the investment options include liquifying local equity, small business/farm aggregation, consumer aggregation, small business incubation, back office and marketing support, debt-for-equity swaps, and the development of community currencies and barter networks.

For more information on the Solari model, please visit http://www.solari.com

Local organizations such as Catherine Austin Fitts' Solari[4] could be established, using the funding provided by local investors to help local citizens out of their financial debt and also helping to bankroll the transition to localized agriculture. These Solaris could even establish local currencies, for use within the local economy. Perhaps it might even be possible to back these local currencies with localized produce. Such a system would allow consumers to escape their debt to corporations, keep our investments in the local community where they can do us the most good, and provide the financing for the necessary transition to relocalization and sustainability.

The road ahead will not be easy, but it will be passable so long as we work together and do not give up. If a grassroots transition succeeds, we may even find ourselves in a better world than we inhabit today.

Resource Guide

The resources listed in this section, including internet links and email addresses, were up to date at the time this book was published.

American Farmland Trust
American Farmland Trust unites farmers, environmentalists, and policy-makers to save America's farmland.
Website: www.farmland.org
CISA: Community Involved in Sustaining Agriculture
A nonprofit organization working to support sustainable agriculture in Massachusetts and throughout the US.
Website: www.buylocalfood.com
Farmer's Market Online
Provides "booth space" for growers, producers, and artisans selling directly to the consumer.
Website: www.farmersmarketonline.com
LocalHarvest
LocalHarvest maintains a definitive and reliable "living" public nationwide directory of small farms, farmers markets, and other local food sources.
Website: www.localharvest.org
Open Air-Market Network
The worldwide guide to farmers markets, street markets, flea markets and street vendors.
Website: www.openair.org

TRANSITIONING
Community Food Security Coalition
The Community Food Security Coalition is dedicated to building

strong, sustainable, local and regional food systems that ensure access to affordable, nutritious, and culturally appropriate food to all people at all times.

Website: www.foodsecurity.org

Earth Pledge

Earth Pledge is a nonprofit organization that identifies and promotes innovative techniques and technologies to restore the balance between human and natural systems.

Website: www.earthpledge.org

The Earth Pledge organization also operates the sustainable localized agriculture website Farm to Table: www.farmtotable.org

EarthSave International

EarthSave is part of a global movement of people from all walks of life who are taking concrete steps to promote healthy and life-sustaining food choices.

Website: www.earthsave.org

Food and Society

Food and Society is an initiative of the W.K. Kellogg Foundation focusing on the development of local ownership and options for economic growth that provide opportunity for maintaining the quality of life most desired by the residents of rural communities.

Website: www.wkkf.org

Food First

The Institute for Food and Development Policy analyzes the root causes of global hunger, poverty, and ecological degradation and develops solutions in partnership with movements for social change.

Website: www.foodfirst.org

Food Secure Canada

Food Secure Canada aims to unite people and organizations working for food security nationally and globally.

Website: www.foodsecurecanada.org

Food Share

Food Share began in the mid-1980s with a mandate to coordinate emergency food services in the city of Toronto. The organization soon began to explore self-help models and cooperative buying

systems, collective kitchens, and community gardens. Their focus has broadened to take in the entire food system, focusing on sustainability, hunger relief, and community involvement.
Website: www.foodshare.net

Foodroutes.org

FoodRoutes is a national nonprofit dedicated to reintroducing Americans to their food — the seeds it grows from, the farmers who produce it, and the routes that carry it from the fields to our tables. Their interactive map lists farmers, CSAs, and local markets across the United States.
Website: www.foodroutes.org

Navdanya

Founded by world-renowned scientist and environmentalist Dr. Vandana Shiva, "Navdanya" means the nine crops that represent India's collective source of food security. The main aim of the Navdanya biodiversity conservation program is to support local farmers, rescue and conserve crops and plants that are being pushed to extinction, and to make them available through direct marketing. Navdanya has its own seed bank and organic farm spread over 20 acres in Uttranchal, in northern India.
Website: www.navdanya.org

Renewing the Countryside

Renewing the Countryside strengthens rural areas by championing and supporting rural communities, farmers, artists, entrepreneurs, educators, activists, and other people who are renewing the countryside through sustainable and innovative initiatives, businesses, and projects.
Website: www.renewingthecountryside.org

Solari

Solari is a model for local investment, founded by Catherine Austin Fitts, that promotes investing in those elements of our economy that protect or invigorate local, living economies.
Website: www.solari.com

COMMUNITY GARDENING/URBAN GARDENING

American Community Gardening Association

The American Community Gardening Association is a nonprofit membership organization of professionals, volunteers and supporters of community greening in urban and rural communities in Canada and the United States. Their website includes excellent information on starting or finding a community garden.

Website: www.communitygarden.org

City Farmer

City Farmer, Canada's Office of Urban Agriculture, publishes *Urban Agriculture Notes,* a great source of information from all over the world.

Website: www.cityfarmer.org

FarmFolk/CityFolk Society

This is a nonprofit society that works with food communities toward a local, sustainable food system. They work on projects that provide access to and protection of foodlands; that support local, small-scale growers and producers; and that educate, communicate, and celebrate with local food communities.

Website: www.ffcf.bc.ca

Growing Power

A very successful organization that involves inner-city youths in community gardening projects, and supports food banks and soup kitchens. This project deserves all the support it can get.

Website: www.growingpower.org

Homeless Garden Project

The Homeless Garden Project employs and trains homeless people in Santa Cruz County, California, within a community-supported organic garden enterprise.

Website: infopoint.com/sc/orgs/garden

The Food Project

The Food Project involves youths and adults in community agriculture and urban gardening. Their services provide healthy food for residents of cities and suburbs while involving volunteers in building their own sustainable food system. Another project that deserves all the support it can get.

Website: www.thefoodproject.org

The Greater Lansing Food Bank Gardening Project

The Garden Project offers support to both home and community

gardeners in the Lansing/East Lansing, Michigan area enabling them to grow and preserve their own fresh vegetables.
Website: lansingfoodbank.org/index.php/garden-project

ORGANIC AGRICULTURE

Canadian Organic Growers

Canada's national membership-based education and networking organization representing farmers, gardeners and consumers in all provinces, COG works with other organizations and government to achieve regulatory change and supports organic events and conferences. Its magazine provides the latest organic news and information, farming and gardening feature articles, plus regular columns. COG's lending library is the only free mail service organic library in Canada. Borrow classic organic texts, periodicals and the latest publications.
Website: www.cog.ca

International Federation of Organic Agriculture Movements

IFOAM is the worldwide umbrella organization for the organic movement, uniting more than 750 member organizations in 108 countries. Its goal is the worldwide adoption of ecologically, socially, and economically sound systems that are based on the principles of organic agriculture.
Website: www.ifoam.org

National Organic Program

The US organic regulatory agency. This is the place to start if you want to protest new standards and regulations.
Website: www.ams.usda.gov/nop/indexNet.htm

National Sustainable Agriculture Information Service

ATTRA — National Sustainable Agriculture Information Service is created and managed by the National Center for Appropriate Technology and is funded under a grant from the United States Department of Agriculture's Rural Business-Cooperative Service. It provides information and other technical assistance to farmers, ranchers, extension agents, educators, and others involved in sustainable agriculture in the United States.
Website: attra.ncat.org

Organic Consumers Association

The Organic Consumers Association (OCA) is a grassroots non-profit public interest organization campaigning for health, justice, and sustainability. The OCA deals with crucial issues of food safety, industrial agriculture, genetic engineering, corporate accountability, and environmental sustainability.

Website: www.organicconsumers.org

Organic Farming Research Foundation

A nonprofit organization whose mission is to sponsor research related to organic farming practices, to disseminate research results to organic farmers and to growers interested in adopting organic production systems, and to educate the public and decision-makers about organic farming issues.

Website: www.ofrf.org

Organic Research Database

A useful database of technical information on organic farming, sustainability issues, and soil fertility. You must be a subscriber to use it.

Website: www.organic-research.com

ORGANIC AND HEIRLOOM SEED

Baker Creek Heirloom Seeds

Preserves the finest in heirloom vegetables, flowers, and herbs.

Address: 2278 Baker Creek Road, Mansfield, MO 65704, USA

Website: www.rareseeds.com

Bountiful Gardens

They sell untreated open-pollinated seed of heirloom quality for vegetables, herbs, flowers, grains, green manures, compost and carbon crops. Specializes in rare and unusual varieties, medicinal plants, and super-nutritious varieties.

Address: 8001 Shafer Ranch Road, Willits, CA 95490-9626, USA

Website: www.bountifulgardens.org

Fedco Co-op Gardening Supplies

Offers a wide variety of untreated seeds and seedlings.

Address: P.O. Box 520, Waterville, ME 04903, USA

Website: www.fedcoseeds.com

Heirloom Seeds
Specializing in heirloom vegetable, flower, and herb seeds.
Address: P.O. Box 245, West Elizabeth, PA 15088-0245, USA
Website: www.heirloomseeds.com

High Mowing Seeds
Another source of certified organic seed.
Address: 813 Brook Road, Wolcott, VT 05680, USA
Website: www.highmowingseeds.com

Irish Eyes – Garden City Seed
Offers organic seed, gardening supplies, and some heirloom seed.
Address: P.O. Box 307, Thorp, WA 98946, USA
Website: www.gardencityseeds.net

Marianna's Heirloom Seeds
Mostly tomatoes and other Italian heirlooms.
Address: 1955 CCC Rd., Dickson, TN 37055, USA
Website: www.mariseeds.com

Organica Seed Company
Provides a large variety of seed that is guaranteed to be grown organically and not subject to genetic modification.
Address: P.O. Box 611, Wilbraham, MA 01095, USA
Website: www.organicaseed.com

Redwood City Seed Company
Heirloom vegetables, hot peppers, and herbs.
Address: P. O. Box 36, Redwood City, CA 94064, USA
Website: www.ecoseeds.com

Salt Spring Seeds
All seeds are untreated, open-pollinated, and non-GMO. They grow all their own seeds and sell only the most recent harvest. A wide selection of grains and beans for the home gardener.
Address: Box 444, Ganges P.O., Salt Spring Island, B.C. V8K 2W1, Canada
Website: www.saltspringseeds.com

Seed Savers Exchange
A nonprofit organization that saves and shares the heirloom seeds of our garden heritage, forming a living legacy that can be passed down through generations.

Address: 3076 North Winn Road, Decorah, Iowa 52101, USA
Website: www.seedsavers.org
Seeds of Change
Organic seed, garden supplies, and some heirloom seed.
Phone: 1-888-762-7333
Website: www.seedsofchange.com
Seeds of Diversity
Canada's premier seed exchange, this is an excellent source of information about heritage seeds, seed saving, and plant diversity.
Address: P.O. Box 36, Stn. Q, Toronto, ON M4T 2L7, Canada
Website: www.seeds.ca
Seeds Trust
Organic, wildcrafted, and heirloom seeds, including native grasses.
Address: P.O. Box 596, Cornville, AZ 86325, USA
Email: support2@seedstrust.com (The McDorman's contend that e-mail has become the most dependable way to contact them.)
Phone Orders: (928) 649-3315
Website: www.seedstrust.com
Seeds West Garden Seeds
Heirloom and open-pollinated, untreated and organic seed. Specializing in seeds that are best adapted for the difficult growing conditions of the west and southwest. Includes many traditional, Native American varieties.
Address: 317 14th Street NW, Albuquerque, NM 87104, USA
Website: www.seedswestgardenseeds.com
Sow Organic Seed
Offers organic and heirloom seed. Provides a lot of useful material on their website.
Address: P.O. Box 527, Williams, OR 97544, USA
Website: www.organicseed.com
Terra Edibles
Organically grown, heirloom seeds.
Address: 535 Ashley Street, Foxboro, Ontario K0K 2B0, Canada
Website: www.terraedibles.ca
The Southern Exposure Seed Exchange
A wonderful source for heirloom seeds and other open-

pollinated (non-hybrid) seeds, with an emphasis on seeds that grow well in the Central Atlantic region. They support seed saving as well as traditional seed breeding.

Address: P.O. Box 460, Mineral, VA 23117, USA
Website: www.southernexposure.com

TomatoFest

Over 500 varieties of organic heirloom tomatoes.
Website: www.tomatofest.com

Two Wings Farm

Two Wings Farm grows gourmet heritage and heirloom vegetables; they do not buy seed from anywhere else for resale. All their seed is certified organic.

Address: 4678 William Head Road, Victoria, BC V9C 3Y7, Canada
Website: www.twowingsfarm.com

Urban Harvest

Urban Harvest provides seeds and garden supplies that promote ecological diversity and preserve the planet's health. All seedlings are grown in or near the greater Toronto area to support the local economy.

Email: grow@uharvest.ca
Website: www.uharvest.ca

PERMACULTURE

Permaculture (Permanent Agriculture) is the conscious design and maintenance of cultivated ecosystems which have the diversity, stability and resilience of natural ecosystems. It is the harmonious integration of landscape, people and appropriate technologies, providing goods, shelter, energy and other needs in a sustainable way. Permaculture is a philosophy and an approach to land use which works with natural rhythms and patterns, weaving together the elements of microclimate, annual and perennial plants, animals, water and soil management, and human needs into intricately connected and productive communities.

— Bill Mollison and Scott Pittman,
La Tierra Community
California
www.permaculture.net

Earth Activist Training

Earth Activist Training blends a full permaculture certification course with Earth-based spirituality, practical political effectiveness, and nature awareness.

Website: www.earthactivisttraining.org

Permaculture Activist

The website for the top print periodical on the subject of permaculture.

Website: www.permacultureactivist.net

Permaculture the Earth

This website features definitions of permaculture along with articles on permaculture design. It also has many links to useful Internet sites and online forums for the permaculture community.

Website: www.permaearth.org

Permaculture Institute

Actively working at ground level on many international and domestic projects, the Permaculture Institute is a nonprofit organization devoted to the promotion and support of the sustainability of human culture and settlements.

Website: www.ibiblio.org/spittman

Permaculture International

Permaculture International Limited's mission is to be a powerful agent for social change towards sustainable, ethical, and cooperative society. It provides services to its members in support of their work in permaculture design and permaculture-related activities.

Website: www.permacultureinternational.org

Permaculture Visions

Permaculture Visions meets the demands of isolated students, working alone, on a tight budget with limited time. It is a leading documenter of permaculture practices.

Website: www.permaculturevisions.com

Tagari Publications

Publishers for the Permaculture Institute.

Website: www.tagari.com

The Central Rocky Mountain Permaculture Institute

The Central Rocky Mountain Permaculture Institute is a permaculture training and counseling research and development cen-

ter. It is part of an alliance of permaculture activists on the cutting edge of agroforestry and other newly discovered methods of organic farming.
Website: www.crmpi.org

The Occidental Arts and Ecology Center

A nonprofit organizing and education center and organic farm in Northern California's Sonoma County. Much of its work addresses the challenges of creating democratic communities that are ecologically, economically, and culturally sustainable in an increasingly privatized and corporatized economy and culture.
Website: www.oaec.org

The Permaculture Research Institute of Australia

The Permaculture Research Institute, headed by Geoff Lawton, is a nonprofit organization involved in global networking and practical training of environmental activists. It offers solutions to local and global ecological problems, and has an innovative farm design in progress. The Institute is also involved in design and consultancy work, and actively supports several aid projects around the world.
Website: www.permaculture.org.au/

The Regenerative Design Institute

RDI emerged from the work of the Permaculture Institute of Northern California and is committed to re-establishing a collaborative connection between humanity and the Earth.
Website: www.regenerativedesign.com/

ECOVILLAGES

Findhorn Ecovillage

The Findhorn Community, which began in 1962 in a caravan park in northeast Scotland, is known internationally for its experiments with new models of holistic and sustainable living. Cooperation and co-creation with nature have always been prime aspects of the community's work, ever since it became famous in the late sixties for its remarkable and beautiful gardens grown in adverse conditions on the sand dunes of the Findhorn peninsula.
Website: www.ecovillagefindhorn.com

The Farm
One of the few hippie communes of the late 1960s and early 1970s to continue to thrive, The Farm is now recognized as one of the most successful ecovillages in the United States.
Website: www.thefarm.org

The Gaia Trust
A Danish-based charitable association for supporting the transition to a sustainable and more spiritual future society through grants and proactive initiatives.
Website: www.gaia.org

The Global Ecovillage Network
An excellent source for finding ecovillages in your region or learning more about ecovillages in general, The Global Ecovillage Network is a global confederation of people and communities dedicated to restoring the land and living "sustainable plus" lives by putting more back into the environment than we take out.
Website: www.ecovillage.org

The Intentional Communities Website
"Intentional Community" is an inclusive term for ecovillages, cohousing, residential land trusts, communes, student co-ops, urban housing cooperatives, and other related projects and dreams. This website provides important information and access to crucial resources for seekers of community, existing and forming communities, and other friends of community.
Website: www.ic.org

SUSTAINABLE AGRICULTURE

Alternative Farming Systems Information Center
Specializes in identifying and accessing information related to alternative agricultural enterprises and crops as well as alternative cropping systems.
Website: www.nal.usda.gov/afsic

Biodynamic Farming and Gardening Association
This is a nonprofit organization formed in 1938 to foster, guide, and safeguard the Biodynamic method of agriculture.
Website: www.biodynamics.com

Ecological Agriculture Projects

Canada's leading resource center for sustainable agriculture.

Website: eap.mcgill.ca/

Farmland Information Center

The FIC is a clearinghouse for information about farmland protection and stewardship.

Website: www.farmlandinfo.org

National Campaign for Sustainable Agriculture

A network of diverse groups whose mission is to shape national policies to foster a sustainable food and agricultural system — one that is economically viable, environmentally sound, socially just, and humane.

Website: www.sustainableagriculture.net

Robyn Van En Center for CSA Resources

Provides a national resource center about CSA for people across the US and around the world. Their "Useful Links" page provides an extensive listing of international, national, state, and regional CSA organizations, as well as Internet resources.

Website: www.csacenter.org

BOOKS

Organic Gardening

All New Square Foot Gardening, Mel Bartholomew, Cool Springs Press, 2006.

Botany for Gardeners, Brian Capon, Timber Press, 1990.

Bugs, Slugs and Other Thugs: Controlling Garden Pests Organically, Rhonda Massingham Hart, Storey Publishing, 1991.

Gardening When It Counts: Growing Food in Hard Times, Steve Solomon, New Society Publishers, 2006.

Growing 101 Herbs that Heal: Gardening Techniques, Recipes, and Remedies, Tammi Hartung, Storey Publishing, 2000.

Growing and Using Herbs Successfully, Betty E. M. Jacobs, Storey Publishing, 1981.

Growing Herbs: For the Maritime Northwest Gardener, Mary Preus, Sasquatch Books, 1994.

Herb Gardening: Why and How to Grow Herbs, Claire Loewenfeld, Faber & Faber, 1964.

How to Grow More Vegetables: And Fruits, Nuts, Berries, Grains, and Other Crops Than You Ever Thought Possible on Less Land Than You Can Imagine, John Jeavons, Ten Speed, 1974.

Let It Rot! The Gardener's Guide to Composting, Stu Campbell, Storey Publishing, 1975.

MacMillan Book of Natural Herb Gardening, Marie-Luise Kreuter, Collier Books, 1985.

Rodale's Chemical-Free Yard and Garden: The Ultimate Authority on Successful Organic Gardening, Miranda Smith, Linda A. Gilkeson, Joseph Smillie, Bil Wolf, Anna Carr (Editor), Rodale Press, 1991.

Save Your Own Seed, Lawrence D. Hills, Abundant Life Seed Foundation, 1989.

Saving Seeds: The Gardener's Guide to Growing and Saving Vegetable and Flower Seeds, Marc Rogers, Storey Publishing, 1990.

Secrets of Plant Propagation: Starting Your Own Flowers, Vegetables, Fruits, Berries, Shrubs, Trees, and Houseplants, Lewis Hill, Storey Publishing, 1985.

The Backyard Homestead: MiniFarm and Garden Log Book, John Jeavons, J. Mogador Griffin, and Robin Leler, Ten Speed, 1983.

The Encyclopedia of Organic Gardening, the Organic Gardening Magazine Staff, Rodale Press, 1959.

The Gardener's Bug Book: Earth-Safe Insect Control, Barbara Pleasant, Storey Publishing, 1994.

The Gardener's Guide to Plant Diseases, Barbara Pleasant, Storey Publishing, 1995.

The Gardener's Weed Book: Earth-Safe Controls, Barbara Pleasant, Storey Publishing, 1996.

The Mulch Book, Stu Campbell, Storey Publishing, 1991.

The One-Straw Revolution: Introduction to Natural Farming, Masanobu Fukuoka, Rodale Press, 1978.

The Ruth Stout No-Work Garden Book, Ruth Stout and Richard Clemence, Bantam Books, 1973.

Tips for the Lazy Gardener, Linda Tilgner, Storey Publishing, 1985.

Winter Gardening in the Maritime Northwest: Cool Season crops for the Year-Round Gardener, Binda Colebrook, Sasquatch Books, 1999.

Permaculture

Gaia's Garden: A Guide to Home-Scale Permaculture, Toby Hemenway, Chelsea Green, 2001.

Introduction to Permaculture, Bill Mollison and Rena Mia Slay, Tagari Publications, 1997.

Permaculture: A Designers' Manual, Bill Mollison and Rena Mia Slay, Tagari Publications, 1997.

Permaculture: Principles and Pathways Beyond Sustainability, David Holmgren, Holmgren Design Services, 2002.

Permaculture in a Nutshell, Patrick Whitefield, Permanent Publications, 1993.

The Permaculture Home Garden: How to Grow Great-tasting Fruit and Vegetables the Organic Way, Linda Woodrow, Penguin Books, 1996.

Ecovillages

Creating a Life Together: Practical Tools to Grow Ecovillages and Intentional Communities, Diana Leafe Christian, New Society Publishers, 2003.

Ecocities: Rebuilding Cities in Balance with Nature (revised edition), Richard Register, New Society Publishers, 2006.

Ecovillages: A Practical Guide to Sustainable Communities, New Society Publishers, 2005.

Ecovillage Living: Restoring the Earth and Her People, Hildur Jackson and Karen Svensson (editors), Green Books, 2002.

Relocalize Now! Getting Ready for Climate Change and the End of Cheap Oil — a Post Carbon Guide, Julian Darley, David Room and Celina Rich, New Society Publishers (in press).

Toward Sustainable Communities: Resources for Citizens and Their Governments, Mark Roseland, New Society Publishers, 2005.

Notes

CHAPTER 1: FOOD = ENERGY + NUTRIENTS

1. P. Buringh, "Availability of Agricultural Land for Crop and Livestock Production," *Food and Natural Resources,* D. Pimentel and C. W. Hall (eds), Academic Press, 1989.

2. P.M. Vitousek et al. "Human Appropriation of the Products of Photosynthesis," *Bioscience* 36, 1986.

3. David Pimental and Marcia Pimental, *Land, Energy and Water: The Constraints Governing Ideal US Population Size,* NPG Forum Series, 1995.
   www.npg.org/forum_series/land_energy&water.htm

4. Henry H. Kindell and David Pimentel, "Constraints on the Expansion of Global Food Supply," *Ambio* Vol. 23 No. 3, May 1994.

5. Mario Giampietro and David Pimentel, "The Tightening Conflict: Population, Energy Use, and the Ecology of Agriculture," NPG Forum Series, 1995.
   www.npg.org/forum_series/tightening_conflict.htm

6. Kindell and Pimentel, "Constraints on the Expansion of Global Food Supply," *Ambio* Vol. 23 No. 3, May 1994.

7. David Pimentel and Mario Giampietro, *Food, Land, Population and the U.S. Economy,* Carrying Capacity Network Publications, November, 1994.

8. Randy Schnepf, *Energy Use in Agriculture: Background and Issues,* CRS Report for Congress, order code RL32677, November, 2004. (Data in the chart from John Miranowski, *Energy Consumption in U.S. Agriculture.*
   www.farmfoundation.org/projects/03-35EnergyConference presentations.htm

9. N.B. McLaughlin et al., "Comparison of Energy Inputs for

Inorganic Fertilizer and Manure-Based Corn Production," *Canadian Agricultural Engineering*, Vol. 42, No. 1, 2000.

10. John Hendrickson, *Energy Use in the U.S. Food System: A Summary of Existing Research and Analysis*, Center for Integrated Agricultural Systems, 2004.
www.cias.wisc.edu/archives/1994/01/01/energy_use_in_the_us_food_system_a_summary_of_existing_research_and_analysis/index.php;
Martin C. Heller and Gregory A Kaoleian, *Life Cycle-Based Sustainability Indicators for Assessment of the US Food System*, The Center for Sustainable Systems, Report No. CSS00-04, 12/6/2000.
www.umich.edu/~css

11. Pimentel and Giampietro, *Food, Land, Population and the U.S. Economy.*

12. FAO, IFAD and WFP, *Reducing Poverty and Hunger: The Critical Role of Financing for Food, Agriculture and Rural Development*, prepared for the International Conference on Financing and Development, Monterey, Mexico, March 18-22, 2002.
www.fao.org/docrep/fao/003/y6265E/Y6265E.pdf

13. Food and Agriculture Organization, *FAO World Summit Progress Report, Sept. 2004*, 2004.
www.fao.org/docrep/meeting/008/J2925e/J2925e00.htm

14. Mark Nord, Margaret Andrews and Steven Carlson, *Household Food Security in the United States, 2004*, Economic Research Service, USDA, Report No. ERR11, October 2005.
www.ers.usda.gov/Publications/err11/

15. Frances Moore Lappé, Joseph Collins, Peter Rosset, *World Hunger: Twelve Myths*, Grove Press, 1998.

CHAPTER 2: LAND DEGRADATION

1. Gretchen C. Daily, "Restoring Value to the World's Degraded Lands," *Science*, 269, pp. 350–354, 1995. Reprinted in L. Owen and T Unwin, eds., *Environmental Management: Readings and Case Studies*, Blackwell, 1997.
Since 1945, soil degradation has affected in excess of 2 billion hectares, or 17 percent of the Earth's vegetated land. Of

the degraded lands, 38 percent are lightly degraded, 46 percent are moderately degraded, 15 percent are severely degraded (reclaimable only with major effort), and 0.5 percent are extremely degraded (unreclaimable). In addition, over 4.5 billion hectares of rangelands are degraded and subject to desertification. The breakdown for rangelands is 27 percent lightly degraded, 28 percent moderately degraded, 44 percent severely degraded, and 1.6 percent extremely degraded. To this survey of land degradation we must also add tropical rain forest degradation, which afflicts more than 427 million hectares. The total of land degraded by soil depletion, desertification, and degradation of tropical rainforests comes to more than 5 billion hectares, or more than 43 percent of the Earth's vegetated surface.

2. R.A. Houghton, "The Worldwide Extent of Land Use Change," *Bioscience* 44(5): pp. 305–313.

3. Ibid.

4. N. Myers, *The Nontimber Values of Tropical Forests,* Forestry for Sustainable Development Program, University of Minnesota, November, 1990.

5. Pimentel and Giampietro, *Food, Land, Population and the U.S. Economy.*

6. David Pimentel et al., "Will Limits of the Earth's Resources Control Human Numbers?," *Environment Development and Sustainability,* Issue 1, 1999.

7. Pimentel and Giampietro, *Food, Land, Population and the U.S. Economy.*

8. Ibid.

9. Ibid.

10. D. Wen, "Soil Erosion and Conservation in China," in *Soil Erosion and Conservation,* ed. David Pimentel, Cambridge University Press, 1993, pp. 63–86.

11. J. R. Parrington et al., "Asian Dust: Seasonal Transport to the Hawaiian Islands," *Science* 246: 1983, p. 195–197.

12. N. Rome Alexandrotos, *World Agriculture: Toward 2010,* Food and Agriculture Organization of the United Nations and John Wiley & Sons, 1995.

13. H.E. Dregne, "Erosion and Soil Productivity in Africa," *Journal of Soil and Water Conservation*, 45, 1990, pp. 431–436.

14. M. Simons, "Winds toss Africa's soil, feeding lands far away," *New York Times*, October 29, 1992. pp. A1, A16.

15. H. E. Dregne, ed., *Degradation and Restoration of Arid Lands*, Texas Technical University, 1992.

16. R. Lal, "Soil Erosion Impact on Agronomic Productivity and Environment Quality," *Critical Reviews in Plant Sciences*, 17, 1998, pp. 319–464.

17. Pimentel and Giampietro, *Food, Land, Population and the U.S. Economy*.

18. Ibid.

19. Ibid.

20. Pimental and Pimental, *Land, Energy and Water*.

21. Pimentel and Giampietro, *Food, Land, Population and the U.S. Economy*.

22. Ibid.

CHAPTER 3: WATER DEGRADATION

1. Sandra L. Postel, Gretchen C. Daily and Paul R. Ehrlich, "Human Appropriation of Renewable Fresh Water," *Science*, 271: 785; Feb. 9, 1996.

2. Sandra L. Postel, *Water for Agriculture: Facing the Limits*, Worldwatch Paper 93, Worldwatch Institute, 1989.

   Sandra L. Postel, "Water for Food Production: Will There Be Enough in 2025?," *Bioscience*, Vol. 48, No. 8, 1998, pp. 629–637.

   According to one study, an estimated 13,800 cubic kilometers of water was consumed in 1995 for food production. This was nearly 20 percent of the total annual evapotranspiration occurring on the Earth's land surface. This translates into an annual average of approximately 2,420 cubic meters per capita. (Evapotranspiration is the combined measure of water evaporation and water uptake by plants and release from plants back into the atmosphere. On a large scale, it is difficult to measure transpiration separately from evaporation, so the two are measured in combination.)

3. Sandra L. Postel, *Water for Agriculture: Facing the Limits,* Worldwatch Paper 93, Worldwatch Institute, 1989.
4. Ibid.
5. Sandra L. Postel, *Dividing the Waters: Food Security, Ecosystem Health, and the New Politics of Scarcity,* Worldwatch Paper 132, Worldwatch Institute, 1996.
6. Postel, *Water for Agriculture.*
7. Pimentel and Giampietro, *Food, Land, Population and the U.S. Economy.*
8. Ibid.
9. Postel, *Water for Agriculture.*
10. H. Frederiksen, J. Berkoff and W Barber, *Water Resources Management in Asia,* World Bank, 1993.
11. Pimentel and Giampietro, *Food, Land, Population and the U.S. Economy.*
12. Ibid.
13. Ibid.
14. Ibid.
15. Postel, *Water for Agriculture.*
16. Pimentel and Giampietro, *Food, Land, Population and the U.S. Economy.*
17. Ibid.
18. Ibid.
19. Ibid.
20. Ibid.
21. Postel, *Water for Agriculture.*
22. Sandra Postel, *Liquid Assets: The Critical Need to Safeguard Freshwater Ecosystems,* Worldwatch Paper No. 170, Worldwatch Institute, 2005.

CHAPTER 4: EATING FOSSIL FUELS

1. Pimentel and Giampietro, *Food, Land, Population and the U.S. Economy.*
2. Pimental and Pimental, *Land, Energy and Water.*
3. Pimentel and Giampietro, *Food, Land, Population and the U.S. Economy.*
4. Ibid.

5. Giampietro and Pimentel, "The Tightening Conflict."
6. Ibid.
7. Ibid.
8. Ibid.
9. Ibid.
10. C. A. S. Hall, C. J. Cleveland and R. Kaufmann, *Energy and Resource Quality*, Wiley Interscience, 1989.
11. Martin C. Heller and Gregory A. Kaoleian, *Life Cycle-Based Sustainability Indicators for Assessment of the US Food System*, The Center for Sustainable Systems, Report No. CSS00-04, December, 2000.
   www.umich.edu/~css
12. Ibid.
13. Susan J. Brown and J. Claire Batty, "Energy Allocation in the Food System: A Microscale View," *Transactions of the American Society of Agricultural Engineers*, 19(4), 1976, pp. 758–761.
14. Assuming a figure of 2,500 kilocalories per capita for the daily diet in the United States, the 10:1 energy ratio translates into a cost of 35,000 kilocalories of exosomatic energy per capita each day. However, considering that the average return on one hour of endosomatic labor in the US is about 100,000 kilocalories of exosomatic energy, the flow of exosomatic energy required to supply the daily diet is achieved in only 20 minutes of labor in our current system.
15. Pimentel and Giampietro, *Food, Land, Population and the U.S. Economy*.
16. Shanthy A. Bowman and Bryan T. Vinyard, "Fast Food Consumption of U.S. Adults: Impact on Energy and Nutrient Intakes and Overweight Status," *Journal of the American College of Nutrition*, Vol. 23, No. 2, 2004, pp. 163–168.
   www.jacn.org/cgi/content/abstract/23/2/163
17. Pimentel and Giampietro, *Food, Land, Population and the U.S. Economy*.
18. Ibid.
19. John Wargo, *Our Children's Toxic Legacy*, Yale University Press, 1998; Ted Schettler et al., *Generations at Risk*, Massachusetts Institute of Technology, 1999.

20. Lennart Hardell and Miikael Eriksson, "A Case-Control Study of Non-Hodgkin Lymphoma and Exposure to Pesticides," *Cancer,* March 15, Vol. 85/ No. 6, 1999.

21. Kristin S. Schafer et al., *Chemical Trespass; Pesticides in Our Bodies and Corporate Accountability,* Pesticide Action Network North America, 2004.

22. Pimental and Pimental, *Land, Energy and Water.*

23. Giampietro and Pimental, Tightening Conflict.

24. Pimental and Pimental, *Land, Energy and Water.*

25. Giampietro and Pimental, Tightening Conflict.

26. Vitousek, "Human Appropriation,"; Giampietro and Pimental, Tightening Conflict.

27. Rich Pirog, Timothy Van Pelt, Kamyar Enshayan and Ellen Cook, *Food, Fuel and Freeways,* Leopold Center for Sustainable Agriculture, 2001.
www.leopold.iastate.edu/pubs/staff/ppp/food_mil.pdf

28. Federal Highway Administration. U.S. Dept. of Transportation, *Highway Statistics 1997.*

29. David Pimentel and Marcia Pimentel, *Food, Energy and Society,* John Wiley and Sons, 1979.

30. Agricultural Marketing Service, U.S. Dept. of Agriculture, 1996.

31 Rich Pirog and Andrew Benjamin, *Checking the Food Odometer: Comparing Food Miles for Local Versus Conventional Produce Sales to Iowa Institutions,* Leopold Center for Sustainable Agriculture, 2003.
www.leopold.iastate.edu/pubs/staff/files/food_travel072103.pdf

32. USDA Economic Research Service, *Import Share of Food Disappearance for Selected Foods, Selected Years,* 2003.
www.ers.usda.gov/data/foodconsumption/datasystem.asp

33. Tim Lang, *Food Safety and Public Health: Will the Crisis Ever End?,* Cardiff Law School Public Lecture Series: 4, Thames Valley University, 2001.

34. Economic Research Service, USDA, 2000, *Foreign Agricultural Trade of the United States.*

35. Stephen Bentley and Ravenna Barker, *Fighting Global Warming at the Farmer's Market,* Foodshare Research Action Report, 2nd edition, Foodshare, 2005.

www.foodshare.net/resource/files/ACF230.pdf

36. Ibid.

37. Giampietro and Pimental, Tightening Conflict.

38. Andy Jones, *Eating Oil: Food Supply in a Changing Climate,* Sustain/Elm Farm Research Centre, 2001.

39. Fölke Gunther, "Phosphorus Flux and Societal Structure," *A Holistic Approach to Water Quality Management: Finding Life-styles and Measures for Minimizing Harmful Fluxes from Land to Water,* Stockholm Water Symposium August, 1992, Stockholm, Sweden, Pub. No. 2, Stockholm Water Co., pp. 267–298.

40. Giampietro and Pimental, Tightening Conflict.

41. Harry Clowes, *Fruit and Vegetable Production in Iowa,* MS thesis, Iowa State College, 1927.

42. Richard Pirog and John Tyndall, *Comparing Apples to Apples: An Iowa Perspective on Apples and Local Food Systems,* Leopold Center for Sustainable Agriculture, 1999. www.leopold.iastate.edu/pubs/staff/apples/applepaper.pdf

43. Giampietro and Pimental, Tightening Conflict.

44. Ibid.

45. Ibid.

46. Ed Ayers, *The History of a Cup of Coffee,* World Watch, 1994.

CHAPTER 5: THE END OF THE OIL AGE

1. Jean Laherrère, *Is USGS 2000 Assessment Reliable?* www.oilcrisis.com/laherrere/usgs2000/

2. "These adjustments to the USGS and MMS estimates are based on non-technical considerations that support domestic supply growth to the levels necessary to meet projected demand levels."

   Office of Integrated Analysis and Forecasting, Energy Information Administration, US Department of Energy, *Annual Energy Outlook 1998 with Projections to 2020,* 1997, p. 221.

3. Colin Campbell, "Country Assessment — Russia," *ASPO Newsletter No. 31,* 2003. www.peakoil.ie/

4. *The Moscow News,* "Russia's Oil Exports Reach Maximum, Decline to Start in Two Years," September, 9, 2004.
www.mosnews.com/money/2004/11/09/oilproduction.shtml

5. Richard C. Duncan and Walter Youngquist, "The World Petroleum Life-Cycle," Presented at the PTTC Workshop *OPEC Oil Pricing and Independent Oil Producers,* Petroleum Technology Transfer Council Petroleum Engineering Program, University of Southern California, Los Angeles, California, October 22, 1998.
www.dieoff.com/page133.htm

6. Andrew Kelly, "US Natgas suppliers seen facing tough challenge," *Reuters,* April 26, 2000.
www.dieoff.com/nagas.htm

7. Charles Esser, *Mexico Country Analysis Brief,* Energy Information Administration, 2004.
www.eia.doe.gov/emeu/cabs/mexico.html

8. Dale Allen Pfeiffer, *The End of the Oil Age,* Lulu Press, 2004.

9. Ibid.

10. Ibid.

11. Ibid.

12. Ibid.

CHAPTER 6: THE COLLAPSE OF AGRICULTURE

1. Pimental and Giampietro, *Food, Land, Population and the U.S. Economy.*

2. Kindell and Pimental, *Constraints on Expansion.*

3. Pimental and Giampietro, *Food, Land, Population and the U.S. Economy.*

4. U.S. Census Bureau, *Poverty 2002,*
www.census.gov/hhes/poverty/poverty02/pov02hi.html

5. Pimental and Pimental, *Land, Energy and Water.*

6. Ibid.

7. Frances Moore Lappé, *Diet for a Small Planet,* Ballantine Books, 1991.
www.dietforasmallplanet.com/

8. Giampietro and Pimental, Tightening Conflict.

9. Ibid.

10. Paul J. Werbos, *Energy and Population*, NPG Forum, 1993. www.npg.org/forum_series/werbos.html;
    David Pimentel et al., "Impact of Population Growth on Food Supplies and Environment," *Population and Environment*, 19 (1): 1997. pp. 9–14.
11. U.S. Census Bureau, *U.S. and World Population Clocks*. www.census.gov/main/www/popclock.html
12. Barbara Tuchman, *A Distant Mirror*, Ballantine Books, 1978.
13. Giampietro and Pimentel, Tightening Conflict.
14. Walter Youngquist, "The Post-Petroleum Paradigm — and Population," *Population and Environment: A Journal of Interdisciplinary Studies*, Volume 20, Number 4, 1999.
    www.mnforsustain.org/youngquist_w_post_petroleum_and_population.htm
15. James H. William, David Von Hippel and Peter Hayes, *Fuel and Famine: Rural Energy Crisis in the Democratic People's Republic of Korea*, Institute on Global Conflict and Cooperation, Policy Paper 46, 2000.
    www.repositories.cdlib.org/cgi/viewcontent.cgi/article=1028&context=igcc
16. D.F. Von Hippel and Peter Hayes, *Demand and Supply of Electricity and Other Fuels in the Democratic People's Republic of Korea*, Nautilus Institute, 1997.
17. Giampietro and Pimentel, Tightening Conflict.
18. Ibid.
19. Tony Boys, *Causes and Lessons of the "North Korean Food Crisis"*, Ibaraka Christian University Junior College, 2000.
    www9.ocn.ne.jp/%7Easlan/dprke.pdf
20. Giampietro and Pimentel, Tightening Conflict.
21. Ibid.
22. Ibid.
23. Jean Laherrère, *Modeling future oil production, population and the economy*, ASPO Second international workshop on oil and gas, Paris, May, 2003.
    www.oilcrisis.com/laherrere/aspoParis.pdf
24. United Nations Development Programme and the UN Food and Agriculture Organization, *DPR Korea: Agricultural Re-*

covery and Environmental Protection (AREP) Program, Iden-tification of Investment Opportunities, Vol. 2: Working Papers 1–3. 1998.

25. Ibid.
26. Youngquist, *Post-Petroleum Paradigm*.
27. Giampietro and Pimentel, Tightening Conflict.
28. Ibid.
29. Ibid.
30. Ibid.

"...the energy cost of ammonia synthesis even in large mod-ern plants averages over 40 GJ/tN, of which 60 percent is feedstock and 40 percent is process energy. It is unlikely that the DPRK fertilizer factories can produce ammonia for less than 50GJ/tN. Further, because ammonia requires special storage and application, most of it is converted to liquid or solid fertilizer (e.g. urea) for easy shipping and application. The conversion of ammonia to urea requires an additional 25 GJ/tN. Since one barrel of oil represents approximately 6GJ of energy, and one ton of nitrogen in urea requires 75 GJ (or more) to produce, to run the DPRK's (three) fertilizer facto-ries at capacity for a year would require:

$$(75 \div 6 = 12.5) \times 400{,}000 = 5{,}000{,}000$$

...or at least 5 million barrels of oil, roughly a quarter of the amount of oil imported annually into the DPRK in recent years."

31. Ibid.
32. Youngquist, *Post-Petroleum Paradigm*.
33. Giampietro and Pimentel, Tightening Conflict.
34. Ibid.
35. FAO, Global Information and Early Warning System on Food and Agriculture, World Food Programme, *Special Report: FAO/WFP Crop and Food Supply Assessment Mission to the Democratic People's Republic of Korea,* November 12, 1998. www.fao.org/waicent/faoinfo/economic/giews/english/alert es/1998/srdrk981.htm
36. Ibid.
37. Youngquist, *Post-Petroleum Paradigm*.

38. Giampietro and Pimentel, Tightening Conflict.
39. Ibid.
40. Ibid.
41. Ibid.
42. Ibid.
43. Ibid.
44. Ibid.
45. Ibid.
46. Ibid.
47. Ibid.
48. Ibid.

CHAPTER 7: THE NEXT "GREEN REVOLUTION":
CUBA'S AGRICULTURAL MIRACLE

1. M. Sinclair and M. Thompson, *Cuba, Going Against the Grain: Agricultural Crisis and Transformation,* Oxfam America Report, June 2001.
   www.oxfamamerica.org/pdfs/cuba/reformingag.pdf
2. P. M. Rosset, "Cuba: A Successful Case Study of Sustainable Agriculture," in *Hungry for Profit: The Agribusiness Threat to Farmers, Food, and the Environment,* ed. F. Magdoff et al., Monthly Press Review, 2000.
   www.foodfirst.org/cuba/success.html
3. Giampietro and Pimental, Tightening Conflict.
4. Sinclair and Thompson, *Cuba, Going Against the Grain.*
5. C. D. Deer et al., "Household Incomes in Cuban Agriculture: A Comparison of the State, Co-operative and Peasant Sectors," in *Development and Change, Vol. 26,* Blackwell, 1995.
6. Giampietro and Pimental, Tightening Conflict.
7. Ibid.
8. Ibid.
9. Ibid.
10. United Nations Development Programme (UNDP), The United Nations Environment Programme (UNEP), World Bank, and World Resources Institute, *World Resources 2000– 2001 — People and Ecosystems: The Fraying Web of Life,*

UDNP, September 2000.
www.wri.org/wr2000/pdf_final/wr2000.zip

11. Ibid.
12. Ibid.
13. Giampietro and Pimental, Tightening Conflict.
14. UNDP et al., *World Resources.*
15. Giampietro and Pimental, Tightening Conflict.
16. Ibid.
17. UNDP et al., *World Resources.*
18. Ibid.
19. Giampietro and Pimental, Tightening Conflict.
20. UNDP et al., *World Resources.*
21. Giampietro and Pimental, Tightening Conflict.
22. Sinclair and Thompson, *Cuba, Going Against the Grain.*
23. Giampietro and Pimental, Tightening Conflict.
24. Ibid.
25. Ibid.
26. Ibid.
27. Ibid.
28. Ibid.
29. Ibid.
30. Ibid.
31. Ibid.
32. Ibid.
33. Ibid.
34. Ibid.
35. Ibid.
36. Sinclair and Thompson, *Cuba, Going Against the Grain.*
37. Ibid.
38. Ibid.
39. Interpress Service, *"Learn from Cuba," Says the World Bank,*
    May 1, 2001.

CHAPTER 8: BUILDING A SUSTAINABLE AGRICULTURE

1. Katherine H. Brown and Ann Carter et al., *Urban Agriculture
   and Community Food Security in the United States: Farming
   from the City Center to the Urban Fringe,* Urban Agriculture

Committee of the Community Food Security Coalition, 2003. www.foodsecurity.org/PrimerCFSCUAC.pdf

2. J. Pretty, J. I. L. Morison and R. E. Hine, "Reducing Food Poverty by Increasing Sustainability in Developing Countries," *Agriculture, Ecosystems and Environment*, 95, No. 1: 2003 pp. 217–234.

3. ADAS Consulting Ltd., *Energy Use in Organic Farming Systems*, Ministry of Agriculture, Fisheries and Food, Project OF0182, DEFRA, London, 2001.

4. John G. Howe, *The End of Fossil Energy and the Last Chance for Sustainability*, 2nd ed., McIntire Publishing Services, 2005. www.mcintirepublishing.com

5. Fölke Gunther, *Vulnerability in Agriculture: Energy Use, Structure and Energy Futures*, presented at the INES Conference, THK, Stockholm, June 15, 2000. www.etnhum.etn.lu.se/~fg/written/stuff/ines/INES.pdf

6. Ibid.

Phosphorus cannot be obtained from the atmosphere by leguminous plants, as is the case with nitrogen. Nor is it as widely available in soils as is potassium, which is supplied by the breakdown of potassium feldspar and various clay minerals. Presently, phosphorus is supplied through the mining of phosphate ores.

Available phosphorus sources may be economically depleted in 130 years at current energy prices. Phosphate ore extraction is an energy-demanding process. Energy depletion may considerably shorten the lifetime of phosphorus mining operations. Incidentally, a valuable amount of phosphates are wasted in detergents and then go unreclaimed through our sewage treatment systems.

7. Ibid.

8. Ibid.

9. Ibid.

10. J. Smit, A. Ratta and J. Nasr, *Urban Agriculture: Food, Jobs, and Sustainable Cities*, United Nations Development Programme, 1996.

11. United Nations Division for Sustainable Development, *Unit-*

ed Nations Sustainable Development Success Stories, Vol. 4.

12. Michael Ableman, "Agriculture's Next Frontier: How Urban Farms Could Feed the World," *Utne Reader,* November–December 2000.

13. R. Heimlich and C. Bernard, *Agricultural Adaptation to Urbanization: Farm Types in the United States Metropolitan Area,* Economic Research Service, USDA, 1993.

14. Interpress Service, *"Learn from Cuba," Says the World Bank,* May 1, 2001.

15. Lang, *Food Safety and Public Health.*

16. Giampietro and Pimental, Tightening Conflict.

CHAPTER 9: TWELVE FUN ACTIVITIES FOR ACTIVISTS

1. Richard Duncan, *The Peak of World Oil Production and the Road to Olduvai Gorge,* C. Pardee Keynote Symposia, Geological Society of America Summit, November 2000. www.hubbertpeak.com/duncan/olduvai2000.htm

2. William Blum, *Killing Hope,* Common Courage Press, 2003.

3. Kim Bartley and Donnacha O'Briain, directors, *The Revolution Will Not Be Televised,* 2003. www.chavezthefilm.com/index_ex.htm

    A detailed review of the movie can be found at www.zmag.org/content/showarticle.cfm?SectionID=45&ItemID=4442

    An interview that includes clips from the movie, is archived at www.democracynow.org/article.pl?sid=03/11/06/1558221

4. www.solari.com/

Index

Page numbers in italics indicate illustrations.

About the Author

Dale Allen Pfeiffer is a novelist, a science journalist, a geologist, and a musician. He is the publisher, editor and main contributor for *The Mountain Sentinel* (www.mountainsentinel.com). Among his other publications are *Giants in Their Steps* (fiction) and *The End of the Oil Age* (nonfiction), and *Collected Poems*. His writings on the subject of energy depletion have been widely praised, and have been referred to in the US Congress, and French and Australian Parliaments. He is also an outspoken activist and a tireless proponent of social and economic equality and direct democracy. He currently resides in the Appalachian region of Kentucky.

If you have enjoyed *Eating Fossil Fuels* you might also enjoy other

BOOKS TO BUILD A NEW SOCIETY

Our books provide positive solutions for people who want to
make a difference. We specialize in:

**Environment and Justice • Conscientious Commerce
Sustainable Living • Ecological Design and Planning
Natural Building & Appropriate Technology • New Forestry
Educational and Parenting Resources • Nonviolence
Progressive Leadership • Resistance and Community**

New Society Publishers

ENVIRONMENTAL BENEFITS STATEMENT

New Society Publishers has chosen to produce this book on Enviro 100,
recycled paper made with **100% post consumer waste**, processed chlorine free, and old growth free.

For every 5,000 books printed, New Society saves the following resources:[1]

13	Trees
1,174	Pounds of Solid Waste
1,292	Gallons of Water
1,685	Kilowatt Hours of Electricity
2,134	Pounds of Greenhouse Gases
9	Pounds of HAPs, VOCs, and AOX Combined
3	Cubic Yards of Landfill Space

---

[1]Environmental benefits are calculated based on research done by the Environmental Defense
Fund and other members of the Paper Task Force who study the environmental impacts of
the paper industry.

For a full list of NSP's titles, please call 1-800-567-6772 *or check out our website at:*

www.newsociety.com

NEW SOCIETY PUBLISHERS